YOU ARE HERE

FROM THE COMPASS TO GPS, THE HISTORY AND FUTURE OF HOW WE FIND OURSELVES

HIAWATHA BRAY

BASIC BOOKS
A MEMBER OF THE PERSEUS BOOKS GROUP
New York

Books published by Basic Books are available at special discounts for bulk purchases
in the United States by corporations, institutions, and other organizations. For more
information, please contact the Special Markets Department at the Perseus Books
Group, 2300 Chestnut Street, Suite 200, Philadelphia, PA 19103, or call (800) 810-4145,
ext. 5000, or e-mail special.markets@perseusbooks.com.

Designed by Trish Wilkinson
Set in 11.5 point Goudy Old Style

A CIP catalog record for this book is available from the Library of Congress.
ISBN: 978-0-465-03285-3 (hardcover)
ISBN: 978-0-465-08098-4 (e-book)

10 9 8 7 6 5 4 3 2 1

To my wife, Dimonika,
who came halfway around the world to end up with me.
Hope it was worth the trip.

Contents

Introduction

I WENT LOOKING FOR A MAP OF BOSTON, AND, PREDICTABLY, I GOT lost. Boston's serpentine streets, many of them unlabeled, are a snare for the unprepared. Luckily, my smartphone soon put me on the right track and brought me to the great Georgian Revival building on Boylston Street that houses the Massachusetts Historical Society. Among the gems of the collection is a tanned and tattered map of Boston, created in 1722 by a British sea captain named John Bonner. Believed to be the first map of the city to be run off on a printing press, Bonner's work and its subsequent revisions would be republished and sold to Boston's residents and transients for the next century or so.

This map would not have helped me find the historical society, or even the land it rests upon; it was salt marsh in Bonner's day and beyond the city limits. Yet it is a detailed and thorough work, displaying the major wharves that supported Boston's seaborne trade, as well as many prominent buildings, including churches, grammar schools, and the "Bridewell"—the city jail. Bonner counted forty-two streets, thirty-six lanes, twenty-two alleys, and three thousand houses. There's no knowing how long he spent in accumulating so much data, but Bonner was about eighty years old when the first edition was published, and he passed away three years later.

Owner of several ships, husband of four wives, and a naval veteran of Queen Anne's War, Bonner would have been accustomed to hard work. So would any mapmaker of that era, when accurate geographical knowledge was so difficult to obtain and the tools for acquiring it so rare and so primitive.

But it is a safe bet that no one will ever work so hard to make a map of our planet, or any part of it. Not too many years ago, great swaths of the world were mapped in only the most primitive, notional sense. Today, the geographical details of nearly every place have been charted in remarkable detail. And many of those places are remapped, thousands of times over, every day, by you and me.

As I read the street map on my smartphone, the map also read me, broadcasting my movements to a database maintained by Google, the giant Internet company that created the phone's software. Google followed me to the historical society. Its server computers timed my footsteps, noted my detours, compared my journey to the archived records of thousands of others heading to the same destination. The phone even listened for the gentle chatter of wireless Internet routers, calculated the location of each transmitter, and then added it to a map almost as precise as the ones generated by satellite radio signals and aerial photographs.

Chances are, by tracking me, Google learned nothing it had not already known. But every previously undetected Internet signal, every untimely traffic jam, adds one more data point to the map. Even the time of your journey matters, for the quickest way to get someplace changes by the day and the hour. So Google's calculation of optimum routes and arrival times is based partly on historical data. How long does it take to get there on a typical Monday at four or Sunday at noon? By collecting fresh data from thousands of smartphones every minute, the directions they give us become ever more precise.

The famed Argentine writer Jorge Luis Borges once told of a kingdom whose cartographers created "a Map of the Empire whose size was that of the Empire, and which coincided point for point with it." A map of such vast size is no more than bizarre, delightful fantasy. But when it comes to detailed, point-for-point representation of the real world, Borges's ideal map is just about within reach. Aircraft and satellites photograph every inch of the world, while sensor vehicles crammed with radars and laser range finders cruise our streets, photographing our front doors and the number of cars in our driveways.

And then there are the location sensors masquerading as portable telephones, carried in the purses and hip pockets of two-thirds of the human race. Those of us who use them are helping to remap the world every day. We are a global band of freelance cartographers, continuing the work of millennia, begun by smarter, tougher people who actually knew what they were doing.

Mapmaking is vital work, but not without risk. The Portuguese Ferdinand Magellan got himself killed before he could complete the first voyage around the globe; it took Henry Morton Stanley three years to trace the Congo River through Africa, and more than half the members of his expedition died along the way.

We casual mapmakers are in no danger of losing our lives; instead, we are at risk of losing our privacy. As they measure the world around us, the devices in our pockets also take the measure of their owners. They create four-dimensional maps that measure our locations not just in space, but also in time. A permanent record of our movements over days, months, and years, these maps can reveal the most salient details of our lives—political and religious beliefs, suspicious friendships, bad habits.

Much of this information is accumulated by corporations like Google and the major cell phone carriers. They mine it for insights about our tastes and habits, the better to target those profitable little ads that pop up on our phones. But in May 2013, a series of revelations by Edward Snowden, a former technician at the US National Security Agency, revealed that American intelligence agencies also scrutinize these databases, enabling them to subject millions of citizens to near-constant location surveillance. Perhaps the NSA's methods have made us a little safer from assault by terrorists, but they have also sown distrust of our own government. Many of us fret that our phones have become battery-powered snitches, betraying our secrets by recording our whereabouts.

Yet few technological marvels have been as marvelous as humanity's victory over the mysteries of location. For most of human history, simply finding our way to some distant place or person was nearly as daunting as planning a trip to the moon. There were no decent maps of the destination or of the land or sea routes leading to it. And even when there were, people lacked reliable tools of navigation to ensure that they were constantly on course. As recently as the nineteenth century, our navigational aids were still remarkably primitive, and great swaths of the planet's surface had yet to be properly mapped.

It all changed in the twentieth century, as the new insights of physics and the demands of war and commerce set loose a torrent of navigational innovation. Radio stations became invisible lighthouses for ships and aircraft. Precisely balanced gyroscopes supplanted the magnetic compass as a direction finder, enabled aircraft to fly straight and level, and led to inertial navigational systems that could guide submarines beneath the North Pole or drop hydrogen bombs on Moscow. A Soviet breakthrough in space exploration accidentally taught American scientists how signals from space could steer ships and planes on earth, while a desperate effort to peek behind the Iron Curtain led to the invention of space cameras capable of generating photographic maps of the entire world. A cash register company developed a new wireless way for computers to chat with one another, and a pair of Boston entrepreneurs used the same technology to map the world's largest cities. And the invention of cheap mobile phones attached digital homing beacons to most of humankind.

This book tells the story of these remarkable accomplishments. We will meet a host of brilliant people, goaded by motives ranging from mercenary to military, whose tireless efforts have given us the superb maps, nearly flawless navigational systems, and tracking technologies that can ensure that we always know exactly where we are.

In Chapter 1 you will find a capsule history of navigation from ancient times to the late 1800s. Some of the most impressive break-throughs happened relatively early, like when a Greek scientist from the third century BC figured out the earth's circumference equipped only with sunlight and a few sticks. We will meet the ancient Polynesian sailors who colonized the islands of the South Pacific, sailing across hundreds of miles of open ocean guided by maps made of seashells and bits of straw.

Chapter 2 brings us into the recognizably modern world of the late nineteenth century. We will see how Guglielmo Marconi and other pioneers of radio realized the medium's value as a tool of navigation. Indeed, radio came along at exactly the right time to solve a brand-new problem—helping aviators find their way through dark and cloudy skies. And the tendency of radio waves to bounce off solid objects led to new location-finding systems that may have played a greater role in winning World War II than the atomic bomb.

Yet radio navigation was not enough. In Chapter 3 we see how engineers like Elmer and Lawrence Sperry and Charles Stark Draper developed self-contained systems that relied on gyroscopes—rapidly spinning disks with a magical tendency to maintain a constant orientation in space. Gyroscopes made possible aircraft that could fly a straight and level course without human intervention and aeronautical instruments that kept pilots on course at night or in bad weather. They also enabled inertial navigational systems that could "feel" every movement of a ship or plane and, from that information alone, calculate its exact position with remarkable accuracy.

The earliest navigators steered by objects that orbited in the heavens. In Chapter 4 a team of American scientists develops a new kind of celestial navigation. Inspired by the radio signals from the Russian Sputnik satellite, they realize that the incoming beeps could be used to steer ships and aim guided missiles. The result of their work is Transit, the world's first satellite navigational system. But it was not the last. Chapter 5 introduces the global positioning system (GPS), Transit's bigger, smarter brother. Born for use in war, this ultraprecise satellite navigational service has reshaped the contours of everyday life for millions.

In Chapter 6 we witness a welcome accident. Millions of homes and businesses have installed Wi-Fi radio networks to connect computers and smartphones to the Internet. Hardly any of them knew that they were also remapping cities around the world. It took a couple of businessmen in Boston to realize that each Wi-Fi device could become a navigational beacon. When they proved the idea worked, they found themselves in a high-stakes showdown with the most powerful technology company on earth.

In Chapter 7 we take to the skies over the former Soviet Union, but not for long. The Russians figured out how to shoot down American spy planes as they attempted to photograph military bases and armament factories. After the downing of a U-2 in 1960, American intelligence was effectively blinded. But just a few months later, the United States had launched a spy satellite, capable of flying far out of range of Soviet missiles while photographing millions of square miles. We will see how this Cold War technology, once top secret, now generates the precise photographic maps we use for planning next year's vacation.

Mapmaking is complex work, best left to professionals. Or is it? With cheap GPS units and Internet-based mapping services, just about anyone can become a part-time cartographer, making corrections and additions to current maps or generating entirely new ones. In Chapter 8 we will see how companies like MapQuest and Google brought cheap digital maps to the masses and then gave us the tools to modify and improve them to our heart's content. And we'll see how do-it-yourself mapmaking has become a vital tool for human rights activists and disaster relief workers.

With our digital maps and GPS-enabled phones, we can find anyplace with ease, but others can also find us. In the final two chapters, we will consider the implications of this new locational transparency. In Chapter 9 we will see how businesses seek to profit from their knowledge of our whereabouts. Mobile marketing companies and popular apps like Foursquare persuade us to share our location data. Then they use it to ping us with precisely targeted advertising messages and tempting special bargains. Meanwhile, companies like Point Inside and Wifarer are tackling the last great challenge in personal navigation, by mapping the vast interior spaces of airports, shopping malls, and other giant buildings.

But not everyone who tracks our location bothers to ask for permission. In Chapter 10 we will see how government agencies ranging from the Federal Bureau of Investigation (FBI) to local school boards monitor the movements of citizens. With GPS trackers in our cars, cell phone records subpoenaed from compliant wireless companies, and cameras that record the license plate numbers of passing cars, the cops can keep tabs on us with almost Orwellian diligence. They may be guided by good intentions, but in the near future government agents may be able to track everybody's location, everywhere, all the time, and at very little cost. What's to stop them?

Preserving our privacy in a world full of location-aware technologies is a challenge we may never completely master. But as we apply ourselves to the task, we now have free run of a planet we have thoroughly mapped and photographed and where even the farthest traveler can count on arriving at his intended destination.

This book is about how we got here.

The Hard Way

On November 8, 1620, Captain Christopher Jones knew exactly where he was for the first time in sixty-five days. The trouble was that Jones and his vessel were in the wrong place, 220 miles north of the mouth of the Hudson River, where they had meant to go. For a day Jones headed south, hoping to make good his navigational error. Shifting tidal currents and the shallow, treacherous coastline made every mile of the journey perilous. And with winter and scurvy setting in, Cape Cod began to look appealing to the captain, crew, and passengers of the merchant ship *Mayflower*. By November 11, Jones had steered the ship around, tucked her into Provincetown Harbor, and began putting the pilgrims ashore.[1]

It's a treasured bit of early American lore. But apart from the historical significance of the *Mayflower*'s voyage, it wasn't all that unusual a tale in the seventeenth century. A brave and daring band of travelers set out on a vast journey with only a general idea of where they would end up, the barest notion of how they would get there, and the vaguest suspicions of their whereabouts at any moment during their travels. It's a wonder anyone ever got anywhere back then. And sure enough, many people never did.

For most of human history, millions lived their lives within a few miles of the places where they were born. There were numerous reasons: poverty, the perils of the journey, the lack of cheap, rapid means of transportation. Even those undeterred by such concerns—merchants, soldiers, explorers, religious pilgrims—faced the substantial problem of figuring out how to get where they wanted to go.

By the time the *Mayflower* set sail, humans had made some progress in solving the problem. Jones, passengers, and crew had set out on a 3,000-mile sea voyage, reasonably confident that they would reach their destination. However, this confidence rested on a fragile foundation. The *Mayflower*'s navigational tools were little more advanced than those a Phoenician sailor might have used in the Mediterranean two millennia earlier.

WHAT DO WE NEED TO MAKE OUR WAY SAFELY AROUND THE WORLD? As visual creatures, we prefer to navigate by sight, even when heading to places well beyond the horizon. So from our earliest days, we have drawn maps—pictures that show the distance between one place and another and the direction we must travel to get there.

To draw such maps, or to follow them, people first need a way to pinpoint their location even as they are moving about. They also need a method to ensure that they are moving in the right direction and a way to track the speed at which they are traveling. Today this sounds like a pretty basic tool kit. Yet it took humans about six thousand years to fill it up, as they fumbled their way around the planet on the road to navigational mastery.

The map was likely the first tool. People may have begun drawing maps even before they had mastered writing. They didn't need advanced navigational tools to map a small area. A knack for making recognizable sketches of familiar objects and landmarks would suffice. African rock carvings of forty thousand years ago depict nomadic settlements and cattle pens. Some consider them maps of a sort, though this notion is open to dispute. In 2009 scientists at the University of Zaragoza discovered that a stone found in a Spanish cave fifteen years earlier had been etched with lines representing nearby hunting grounds and a river. The researchers estimated that the etchings had been made nearly fourteen thousand years ago, making it the oldest Western European map uncovered so far.[2]

Later maps were created with more specific uses in mind. A map carved into a clay tablet circa 2300 BC and found in northern Iraq features images of towns, rivers, and mountains, as well as written notes on the dimensions of various plots of farmland. Other maps from the same era were clearly used for tracking landownership and usage; there is even evidence of surveyors' notes, used to resolve boundary disputes and levy taxes.

Maybe the most remarkable early maps were the ones created by the people who colonized the Pacific Ocean. Beginning in Southeast Asia, the Lapita people island-hopped their way to Taiwan, the Philippines, and the Bismarck Archipelago. By 1000 BC, Lapita voyagers had traveled as far east as Fiji and Samoa. Their descendants, the Polynesians, continued the oceangoing tradition, gradually finding their way across the Pacific. By the first millennium AD, the Polynesians had settled in New Zealand, Easter Island, Hawaii, and other islands, separated from one another by thousands of miles of blank blue water. These voyages, unrecorded and unrecalled, rank among the greatest of human explorations.

How did they manage such remarkable journeys without so much as a compass? Sailors relied on nature, reading the subtle but plentiful signs of wind, waves, and wildlife. Anthropologist David Lewis, who worked with modern Polynesians trained in traditional navigational techniques, witnessed their ability to steer in part by the rolling of the sea. Years of experience had taught them to sense their location and direction by feeling the motions of the waves.[3]

Residents of the Marshall Islands recorded this profound knowledge of the sea in three-dimensional maps of sticks and shells. The shells represented islands, while the shape and placement of the sticks illustrated the wave patterns in nearby waters. A sailor would place the stick map on the deck of his boat and then steer until the stick patterns matched the shapes of the waves. More than with other maps, perhaps, the utility of the wave map was dependent on the navigator's knowledge and skill in matching representation and reality.[4]

Apart from the navigational ingenuity of Pacific Islanders, for most of history the development of reliable maps was severely hindered by the lack of trustworthy navigational tools. Accurate mapping requires knowing where you are relative to other places whose exact location is also well known. It would take our species thousands of years to master the craft.

But that doesn't mean we stayed in place. Travelers developed a variety of methods for feeling their way along. They memorized the sea currents and prevailing wind patterns. They followed fellow travelers—fish, turtles, or seabirds—that were heading toward the same lands they sought. Some Norsemen traveled with birds, which they would release from time to time. If the bird returned, they sailed on. If it didn't, they would aim their ship in the direction the bird had flown, fairly confident that they would strike land. If this tactic sounds familiar, you are thinking of the biblical Noah, who launched avian emissaries from his ark.[5]

Of course, like today's travelers, the ancients were guided by signals from space. People had long noticed the regular, predictable motions of the sun, moon, and stars. Even the most ignorant saw that the sun always appears from roughly the same direction, sweeps across the sky, and disappears in a roughly opposite direction. Indeed, the sun's motion probably gave humans their first firm concepts of East and West. The sun also helped define North and South: in the Northern Hemisphere, where most of the human race resides, the sun is usually south of an observer, and due south at solar noon, when the sun is at its highest point in the sky.

At night, travelers relied on the light of more distant suns. Centuries of observation had revealed the predictable movements of stars. In time, this understanding enabled fairly accurate measurements of a traveler's latitude—his position north or south of the equator.

Precise stellar navigation probably began with Egyptian scholars of the third millennium BC, upon recognition of a blessed coincidence.

Each year's flooding of the Nile began on the same day that a familiar star—Sirius, the Dog Star—rose early in the morning, just before dawn. It happened every year, like clockwork. Indeed, this event was a timepiece of sorts, accurate enough to lead the Egyptians toward the development of twenty-four-hour days and the 365-day calendar we use today. Equipped with their naked eyes and an ample supply of darkness, Egyptian astronomers gradually compiled charts of thirty-six constellations. They learned when particular stars would appear at particular points in the sky. And with that knowledge, they took the first great step toward a rigorous science of location.

Close observation of the movement of constellations revealed relatively fixed points that could be reliably located, regardless of the day. For instance, astronomers noticed that some constellations never fall below the horizon; they appear to wheel overhead in an endless circle. Near the axis of that vast wheel, ancient observers could clearly see the Great Bear and the Lesser Bear—what we now call the Big and Little Dippers. One star in the Little Dipper, called Kochab, seems to hang at the center of the circle. For navigators, this became the polestar, the indicator of true north.[6]

Yet adding to the complexity of navigation was the fact that the great stellar wheel is a bit off-kilter. The gravitational pull of the sun and moon causes axial precession, a slow back-and-forth wobble in our planet's orbit. Even thousands of years ago, educated people knew that the earth is a sphere, and as far back as 450 BC, the Greeks figured out that the planet tilts on its axis by about 23 degrees. It is this tilt that causes our seasons, as the earth's hemispheres lean away from the sun and then toward it. In addition, the planet is not a perfect sphere; it is a bit thicker and more massive at the equator. As the planet tilts on its axis, the sun and moon tug unevenly on that extra mass in the middle. The result is precession, the slow wobble first noticed by the Greek astronomer and mathematician Hipparchus around 130 BC.[7] Thanks to precession, the star hovering closest to true north has gradually changed over the centuries, in a cycle that repeats itself every 26,000 years. Although Kochab was the North Star for the ancients—and will be again someday—in our own time, the polestar is Polaris, which is actually a trio of stars about 430 light-years away.

Give a sailor a polestar and some basic trigonometry, and he can figure out how far north or south of the equator he is, using the measurement we call latitude. He would simply assume the polestar is directly above the North Pole and then measure the angle between the star and the horizon. Sailors did this using various devices. There was the astrolabe, a simple machine made first of wood and later of metal, which accurately measured celestial angles. Yet simpler methods also worked. One popular tool was a knotted string held between the teeth and extended toward the sky. If the angle between the polestar and the horizon was, say, 40 degrees, then you were at 40 degrees north latitude, the same as Philadelphia.

By the Middle Ages, European sailors were traveling south of the equator. There, the north polestar was invisible, hidden by the bulk of the planet. And there was no corresponding star hovering over the South Pole. The closest substitute was a constellation called the Southern Cross, and sailors learned to use it as a guide to southern latitude. In addition, they relied on measurements of the sun's angle above the horizon. However, the earth's axial tilt constantly alters the sun's position relative to the equator in a process called declination. As a result, the sun's path through the sky shifts every day of the year. Could so changeable a star offer a reliable guide to latitude?

In fact, the answer is yes. It's possible to determine latitude by measuring the angle of the sun above the horizon at midday. To do it accurately at any spot on earth, however, one needs a declination table—a reference book that lists the sun's angle relative to the equator on that particular day. In 1473 Abraham Zacuto, a Spanish Jew of remarkable erudition, began work on a set of solar declination tables. It took him five years, and the finished work wasn't widely published until 1496. But this book, the *Almanach Perpetuum*, appeared just in time for Europe's age of discovery. Solar declination tables enabled any skilled seaman with a clear view of the sun to determine his latitude, on any ocean, anywhere.[8]

An understanding of latitude was essential to navigation, as was the creation of reliable maps. Yet it was not enough to know how far north or south one was. Precise navigation required a way of accurately measuring one's east-west location—the problem of longitude.

By the second century BC, Greek mathematician Hipparchus had suggested that a proper map of the earth would feature lines of latitude running east and west and parallel to each other, from the equator up and down to the poles. These would be intersected by longitudinal lines running north and south, although not in parallel. Instead, these lines would be farthest apart at the equator and then steadily converge until they united at the poles. Viewed this way, the planet was wrapped in lines, and any place on earth could be defined by its location on the grid.

Three centuries later, the greatest cartographer of antiquity based his maps on the same concept. Claudius Ptolemy's *Geography*, written in the mid-second century AD, set a standard of excellence that would not be improved upon for more than a millennium. Sadly, Ptolemy's original maps have been lost to us. In the centuries before the printing press was invented, books like Ptolemy's were copied by hand, but few copyists had the skill to reproduce maps. Indeed, Ptolemy's maps and text alike were lost to Europe after the collapse of the Roman Empire. Muslim scholars translated the text into Arabic during the ninth century AD; in the thirteenth century, Byzantine monk Maximus Planudes translated it into Latin and reintroduced the book to the Western world. Scholars soon re-created the lost maps using the facts and figures in Ptolemy's text.[9]

Few books have had more impact on history. During the Middle Ages, European maps were works of theology and mythology rather than geography. They purported to show the precise location of the Christian paradise or the kingdom of Prester John, an imaginary hero who might someday aid Christendom in its war against Islam. Ptolemy's maps, by contrast, were an ice-water slap to Europe's dozing geographers. They still have the power to startle. With their latitudinal and longitudinal grids, their cool, unimaginative focus on reality, these reproductions of two-thousand-year-old data look surprisingly modern. Indeed, they are like electronic circuit diagrams compared to the crudities of the medieval cartographers. However imperfect, Ptolemy's maps bear a general resemblance to the real world.

That said, Ptolemy's details were often wrong. Most famously, he grossly underestimated the circumference of the earth—a beneficial blunder, it turns out, as it helped convince Christopher Columbus that he could easily reach Asia by sailing west. But for all his mistakes, Ptolemy laid out the guiding principles for a truly scientific cartography.[10]

By this time, sailors had mastered a simple, almost magical, tool that let them determine their direction of travel. But the magnetic compass—ubiquitous now—was a long time coming. The naturally magnetic rock called lodestone had puzzled scholars for centuries. The earliest written mention of it was by Greek philosopher Thales of Miletus around 585 BC. There is also a curious Chinese account from the second century BC, which tells of the palace of emperor Ch'in Shi Huang Ti. The main gate of the palace was made of a huge lodestone that exerted such great force that men armed with iron weapons could not enter without being detected. By the first century AD, Chinese writers told of ladles made of lodestone and used in mystical rituals meant to foretell the future. No one can vouch for the quality of advice these large spoons provided, but they were certainly consistent: for some reason, the handles of these curious ladles always pointed toward the south. The ancients didn't recognize that the lodestone's magnetic field was lining up with a far more powerful field generated by the entire planet.

A lodestone will share its remarkable power with any piece of iron. Just stroke the iron with the lodestone, and it too becomes a magnet. This parlor trick took on real value when performed on a piece of iron small enough to float. In 1040 AD a Chinese text describes a paper-thin leaf of magnetized iron, shaped like a fish, that could float in a bowl of water. The head of the fish would rotate until it faced south, with the tail aiming to the north. By 1111 another Chinese book describes a magnetized needle that always pointed south, and thus served as a reliable aid to navigators. It is the earliest known reference to the compass, a direction-finding technology that would make its way to Europe late in the twelfth century.

Armed with a compass, sailors could identify North and South, East and West, without reference to heavenly or earthly landmarks. The compass dramatically reduced the peril of long voyages. Prior to its widespread use, sailors kept close to port in the winter months, for fear of storms that would prevent navigation by the stars. The compass enabled a navigator to know which way he was going, no matter the weather.

Along with a compass and an astrolabe or other latitude-measuring tool, a sailor could employ a chip log, a piece of wood bound with knotted rope that was tossed overboard at regular intervals. The rate at which the knots slid through the sailor's hand gave him a reasonably accurate measure of his ship's speed.

Sailors also relied on the sounding line, an ancient tool of great value when navigating relatively shallow water. A knotted rope with a heavy weight at one end was tossed into the sea. The base of the weight was hollowed out and packed with grease or soft wax. When a sailor felt the sinking weight hit bottom, he would count the knots to measure the depth of water beneath the ship's keel. Then he would pull up the rope and examine the weighted end. Bits of the ocean floor would be gummed to its greasy bottom—gravel, perhaps, or sand or seaweed. Over the centuries, thousands of such soundings enabled chart makers to accurately show the depth of water at various points along a coastline and the material to be found on the bottom. A sailor could confirm his ship's position against the chart by doing soundings of his own.

With these tools, a sailor who knew his business could travel a long way through stormy weather while retaining a fair idea of where he was. Starting from the last accurate celestial fix, he would take frequent compass and chip-log readings, noting every variation in course and speed. From these, he could deduce his probable location. This process is called "dead reckoning," and, as I'll show later, a far more sophisticated version of the method that still guides jumbo jets across the Pacific.

Despite explorers' reliance on the compass, for the longest time, nobody knew why magnetized iron pointed to the north. The great sixteenth-century mapmaker Gerhard Mercator, like other wise men of his time, believed in the existence of Rupes Nigra, a vast black island at the top of the world, made wholly of lodestone. The idea was that only so great a magnet could attract all the world's compass needles to itself. It wasn't until 1600, six years after Mercator's death, that an English physician named William Gilbert finally grasped the truth.

Gilbert, who was appointed physician to Queen Elizabeth I the following year, made for himself miniature copies of the earth, carved from lodestones. He found that when moving a compass over these "earthkins," the needle behaved just as it would toward the real earth. For instance, the needle tended to be drawn to the "north" of the earthkin when held parallel to the magnetic globe, but when held point downward, the needle would exhibit "dip," a curious occurrence recorded by another Englishman, Robert Norman, in 1581. Norman had found that the magnetic field tugged the needle downward as well as toward the north. His superbly balanced and calibrated needle pointed not straight down, but consistently at an angle of about 71 degrees.

Gilbert's dip experiment with his earthkins produced similar results, albeit on a much smaller scale. Yet Gilbert went one step further, noting that the downward angle of his needle altered when held over different parts of the earthkin. The farther north the needle was placed, the more its northern end would dip. Move to the south of the earthkin, and the southern end would start dipping. And at the equator, there would be no dipping tendency at all. Confronted with such evidence, Gilbert came to a remarkable and essentially correct conclusion: there was no giant magnet at the North Pole. The giant magnet, Gilbert declared, was the earth itself. It was the first step toward a scientific understanding of the planet's magnetic field.[11]

The compass became a vital tool for mapmakers. A ship's navigator, sailing close to shore, could take compass bearings on key landmarks to accurately fix their relative positions. By compiling enough such readings taken from a variety of locations, cartographers were able to make vast improvements in the accuracy of coastal maps. Over time, mariners' charts stopped depicting shorelines as indistinct slabs of terrain. Instead, these charts revealed every feature of the shore—bays, river deltas, sandbars, cliffs, beaches. As people became better navigators, they learned to produce better maps.[12]

Yet even with better technology, mapmaking is at best an exercise in approximation. Even in Ptolemy's time, mapmakers understood they would have to make compromises with reality. After all, a map seeks to depict the features of a sphere on a flat surface; any such projection must introduce some distortion. Ptolemy and his heirs developed a variety of projection methods, each designed to accurately display some aspect of geometry, while inevitably misrepresenting others. For example, a world map projection that accurately displays the relative sizes of the continents can prove unreliable for calculating the distances between landmasses.

By the sixteenth century, as Europeans set out into the Atlantic to exploit newly discovered lands, what they most needed were maps that made it easier to chart an accurate course from one place to another. On an ideal map, a navigator would be able to draw a straight line from the island of São Miguel in the Azores to the town of Touros in Brazil and know that he would reach his destination by sailing along the compass heading indicated by that line. However, there's no such thing as a straight line, not when you're traveling twenty-six hundred miles over the surface of a sphere. The navigator's ship would actually be sailing in a curve, a far more difficult course to draw on a flat sheet of paper—unless someone could design a map projection that would accurately represent the compass bearing as a straight line.

German-born Gerhard Mercator accomplished this feat in 1569, with a projection that made him history's most famous mapmaker. In his projection, the lines of longitude ran north to south, with each line being equidistant from the others. The latitudinal lines were exactly perpendicular, running east to west. But to correct for the earth's curvature, Mercator drew his latitudinal lines at varying distances from each other; the farther north or south from the equator, the wider the distance between them.

This approach to mapmaking has some disconcerting side effects. Landmasses near the equator are represented with fair accuracy, while those nearer the North and South Poles are greatly distorted, so that their relative sizes are exaggerated. For example, Mercator's map makes Greenland look far too large relative to Europe. No matter; Mercator had designed a map for navigators. His distortions may have made the continents look funny, but they greatly simplified the task of plotting an accurate course at sea with map and compass.[13]

Still, the compass was a fickle guide. Any halfway competent navigator could see that the magnet's idea of North often didn't always line up with the observed position of the polestar. That's because the planet's magnetic field is not uniform. It alters depending on where one is located, in a phenomenon called magnetic declination. Sailors soon learned not to rely too heavily on the compass alone, but to cross-check its readings against traditional methods of stellar navigation. Yet they suspected that there was a rhyme and reason to such compass errors—and that decoding it would eventually lead to a solution to the previously intractable problem of east-west location: longitude.

Some thought that the key had been discovered in 1544. That year, Venetian explorer Sebastian Cabot, who sailed under the flag of Spain, issued a map that showed a spot in the ocean, near the Azores, where magnetic compasses showed no declination at all. Here, the compass needle agreed with the polestar and pointed toward true North. Many a sailor rejoiced at the news, believing that magnetic declination would provide an easy way to measure longitude.[14] They assumed that as a ship sailed east or west, magnetic declination would change in a smooth, regular manner. If travelers on land or sea made careful measurements of true North and magnetic North at many locations around the globe, they would eventually produce a magnetic declination table that would show navigators how far off their compasses would be at any given point. More important, the same table would amount to a reasonably accurate map of longitude. One French navigator declared that each degree of magnetic declination would equal 22.5 leagues—about 56 miles. Thus, you could measure your travel east or west by simply counting off degrees of declination on your compass.

Navigators soon realized that it wasn't quite that simple. Declination didn't change in a predictable east-west fashion. Indeed, navigators found that the compass became ever more erratic as a ship moved farther north. Still, it might be possible to record these variations at many points on land and sea and thus create a magnetic chart of longitude.

In 1698 Queen Mary II of England bankrolled the effort to create such a chart. The royal treasury financed the construction of a small vessel and chose eminent mathematician and astronomer Edmond Halley to lead the expedition. Over two years and two voyages, Halley and his team carefully collected magnetic data over a vast area of the Atlantic Ocean, from 52 degrees north of the equator to 52 degrees south. The result was the first magnetic declination chart.

Halley's chart, and many that have been created since, is a superb resource for navigators, enabling them to correct their compass readings. But Halley soon saw that magnetic declination was useless as a guide to longitude. The reason was something first noted in 1634 by another English mathematician, Henry Gellibrand. Not only did the magnetic field vary from place to place, but the amount of variation changed over time. Using historical records, Gellibrand found that the amount of declination in London had changed from 11.3 degrees east in 1580 to just 4.1 degrees east in his own time.[15] Today, we know that declination changes because of the constant alteration of the planet's magnetic field. That field is generated by a vast ball of molten iron at the planet's core. The nonstop movement of this liquid core, stirred by its own heat and the rotation of the planet on its axis, accounts for the variations in the magnetic field. Indeed, people understand the process well enough to accurately predict the correct magnetic declination at various locations years in advance.

Halley knew none of this, but soon realized that Gellibrand had been right. Whatever the cause of the earth's magnetism, its constant fluctuation ensured that any chart of magnetic declination would have to be regularly updated to be of any use. And worse, such a chart could never provide a means to calculate longitude.

The maddening thing was, every navigator knew how to solve the latitude-longitude problem. It was simple, really—if you had the right tools.

The earth is a near sphere that makes a complete rotation on its north-south axis once every day. Thus, our planet rotates 360 degrees each day, or 15 degrees in each of a day's twenty-four hours. This means you can figure out your east-west location relative to some reference point on land, if you can answer two simple questions. First, what time is it where you are? Second, at that same moment, what time is it at your land-bound reference point?

A navigator could tell local time from the position of the sun, realizing that it was, say, nine in the morning at his location. Now if he knew that it was noon at some point on land, he would know that he was 45 degrees west of that point. How far is that in terms of miles? It depends how far north or south you were. At the equator, 15 degrees of longitude works out to about 1,035 miles of distance, or 69 miles for each degree. But the lines of longitude converge at the North and South Poles, so the distance covered by 15 degrees of longitude gets smaller as you move north or south of the equator. Luckily, a simple formula can be used to calculate the distance, as long as you already know your latitude. And, of course, navigators had already mastered latitude. Thus, if a sailor could nail down the calculation of longitude, he could quickly and accurately figure out his position on any ocean.

So what the sailor needed was a highly accurate timepiece that could operate for months without gaining or losing more than a few minutes. It must be undaunted by constant exposure to saltwater, dramatic variations in temperature and humidity, and the constant pitching and rolling of a ship at sea. In short, our ancient mariner needed a good waterproof wristwatch, the sort we can buy any day of the week at the nearest Walgreens. But the primitive pocket time-pieces of the sixteenth century weren't nearly accurate or durable enough for seagoing navigation. This didn't deter the explorers and empire builders of Europe; the beckoning wealth of the New World was irresistible. Still, solving the longitudinal problem would make their voyages faster, safer, and more profitable.

At first, scientists looked to the heavens for an answer. When in 1610 Galileo discovered four brilliant moons orbiting the planet Jupiter, he proposed using them as a universal clock. First, Galileo pains-takingly recorded the movements of the moons relative to dates and times on earth. By 1612 he had put together a record that enabled him to tell the time of day simply by viewing through a telescope the positions of the Jovian moons. And because the moons would appear the same to any earthbound observer, their position could be used to tell the time in a distant city—Rome, say, or London or Paris.[16]

It was a brilliant idea, if you happened to be on land. But try peer-ing at the moons of Jupiter through a telescope on the rolling deck of a ship at sea. It was a miserable, frustrating business. Besides, Galileo would soon find himself answering to the Catholic Inquisition for un-orthodox views on astronomy. No wonder his method never caught on.

A more practical heavenly clock could be found much closer than the moons of Jupiter. In 1514 a Nuremberg priest named Johannes Werner suggested that longitude could be calculated by measuring the exact position of the earth's own moon. A table showing the distances between the moon and other celestial bodies on every day of the year would be a universal clock, just like the positions of the Jovian moons, but a lot easier for sailors to view.

It took well over two centuries for someone to make a serious at-tempt at creating it. In 1714 the government of Great Britain began offering enormous cash prizes—worth millions of today's US dollars—to anyone who could find a practical way to calculate longitude at sea.

The prize went unclaimed for decades—a measure of the problem's difficulty.

In 1761 a British Anglican cleric named Nevil Maskelyne took up the challenge, basing his effort on Werner's work. In 1766 he published tables showing the relative positions of the sun, moon, planets, and various stars, as viewed from Greenwich, England, for every day of the year 1767. A sailor thousands of miles away could now calculate the time at Greenwich by using a sextant to measure lunar distance and then compare his results to Maskelyne's tables. The lunar tables prepared by Maskelyne and his colleagues formed the core of a new nautical almanac. Published every year, it became an indispensable guide to the world's navigators. Eventually Greenwich would become the baseline used by sailors worldwide to calculate longitude.

However, the lunar method wasn't nearly accurate enough for the remarkable man who finally solved the problem for good. British carpenter and self-taught clockmaker John Harrison spent about forty of his eighty-three years of life designing timepieces that could withstand the stresses of a sea voyage while still keeping accurate time. His H4 marine chronometer of 1755 was a near miracle of engineering, losing just a few seconds of accuracy even after months at sea.

Harrison's invention was just what the world's navigators had needed. Still, most would do without for another century or so. Like early mainframe computers, marine chronometers were immensely expensive. Many navigators continued using the lunar method into the mid-nineteenth century, when accurate ship clocks at last became relatively cheap.

While navigators grappled with the longitudinal problem, cartographers were figuring out how to accurately map vast areas of land. They eventually hit upon a solution familiar to the ancients, the geometric concept of triangulation. This method allows you to calculate the distance to a remote landmark and its latitude and longitude, without having to set foot in the place. You start with a line on the ground, its length very precisely measured and its exact position known. Then you set up some kind of observation device at both ends of the line and aim it at your landmark. If you measure the landmark's angle relative to both ends of your original line, you now have two angles and the length of your line. With these three bits of data, you can calculate the distance to the landmark, as well as its position on the earth's surface.

In the late seventeenth century, Italian geographer Giovanni Cassini moved to France and, with the support of King Louis XIV, began triangulating his way across the countryside in the first systematic effort to map an entire nation. The effort required more than a century and the skills of three generations of Cassinis. Yet when it was completed in 1793, the map of France had set a new standard in geographic precision that other nations would soon emulate.[17]

From the eighteenth century on, the scientific mapping of the planet proceeded in earnest. In the Pacific Ocean, British seaman James Cook charted the coast of New Zealand and the east coast of Australia. In India, British soldiers and explorers mapped the subcontinent in a grueling effort that lasted nearly 150 years. In North America, a young George Washington surveyed the hills and valleys of colonial Virginia; later, Thomas Jefferson, third president of an independent United States, sent Meriwether Lewis and William Clark on a quest to map the new nation's interior. Little by little, humans drew trustworthy pictures of the world.

Not every mapmaker was on the same page, though. Different nations used different methods and standards to draw their maps. As a result, even by the late nineteenth century, it was still impossible to stitch together a consistent, coherent world map. In 1891 a German geographer named Albrecht Penck proposed to unify cartographic efforts. At an international conference of geographers held in Bern, Switzerland, Penck proposed the "International Map of the World," a project to map the entire globe using standardized symbols, colors, and, above all, scale—the ratio of a distance on the map to the corresponding distance on land or sea. For instance, given a map with a scale of 1:24,000, 1 inch on the map equals 24,000 inches, or 2,000 feet, in the real world. Such a map would show a vast amount of detail. However, given the tools and techniques of the era, it was quite impractical. Penck suggested a more manageable scale of 1:1,000,000. On such a map, 1 inch equals 1 million inches, or 15.8 miles, on the ground. Geographers dropped Penck's rather grandiose name and took to calling it the Million Map.

Despite the plan's obvious appeal, it languished for years. Work finally began in the ominous year of 1913. The project continued in fits and starts through two world wars and a global depression. It faded away entirely in 1987 after the United Nations, which had taken responsibility for the Million Map after World War II, finally washed its hands of the matter.

By then governments and businesses worldwide had mapped, charted, and photographed the entire planet. Many parts of the world still haven't been mapped to the level of detail that Penck had hoped for, but all of it has been set down with a degree of precision that the captain of the *Mayflower* could scarcely have imagined. As for navigation, we have grown so good at it that getting truly lost is nearly impossible.

As the twentieth century began, humans could confidently find their way to nearly every part of the world. To a world traveler circa 1900, it may have seemed that the problem of navigation had been solved for all time. Certainly, the difficulties that vexed Christopher Columbus or the captain of the *Mayflower* had been decisively resolved. But the new century would bring a host of new navigational challenges every bit as daunting.

How do you navigate a ship that travels not on the surface of the sea, but five hundred feet below it? How do you steer a craft as it soars thirty thousand feet above the earth, with enough fuel to reach its destination and just one chance to get it right? How do you transmit an urgent message to one person, when that person could be anywhere on the planet? New technologies of travel and communications, and the deadly necessities of warfare, have created many such problems, as well as the tools to solve them. We will look first at the most vital of these tools—the invisible electromagnetic beacons that guide our aircraft and ships and footsteps, and sometimes our deadliest weapons.

The New Wave

On June 19, 1944, James Van Allen stood on the bridge of the American battleship USS *Washington*. Above him flew a wave of Japanese bombing planes that were doing their best to kill him. They came close—so close that Van Allen glimpsed the face of one of the attacking Japanese pilots, moments before his plane and the others were ripped apart, their remains plunging into the Pacific.[1]

Van Allen had been as eager to shoot down those Japanese pilots as they had been to attack him, but he stubbornly stayed alive until 2006. In ninety-one productive years, Van Allen would discover the radiation belts around our planet that bear his name and oversee two dozen unmanned space missions, including trips to Venus, Jupiter, and Saturn. But none of his efforts had a greater impact than the technology that saved his life in 1944.

Van Allen and the US Navy had introduced the Japanese to history's first "smart" weapon, an antiaircraft shell that did not have to hit its target to destroy it. Instead, the shell carried a miniature radio system that could determine the location of a nearby enemy aircraft. When the plane came within lethal range, the shell blew itself up, shredding the aircraft with shrapnel. At the Battle of the Philippine

Sea, the proximity fuse savaged the Japanese air force, saving hundreds, perhaps thousands, of American sailors.

After millennia of fumbling efforts, the sciences of location and navigation reached near perfection in the twentieth century. A host of new technologies enabled people to easily find their way through the world and locate practically any object. Of these innovations, none was more important than radio, the transmission of electromagnetic waves through space.

Radio's pioneers were primarily interested in a faster, cheaper way of communicating over great distances. However, they realized its potential as a navigational tool. Like the beacon of a lighthouse, a radio signal could serve as a reference point, an invisible landmark that could guide ships and airplanes to safe harbors. They foresaw that radio waves, like light, would bounce from reflective surfaces, sending echoes of the original signal through the air like ripples in a pond. A radio receiver that captured those echoes would be able to pinpoint the location of a remote object, whether a commercial airliner on final approach or a Japanese bomber attacking an American ship.

Later, as we learned to build satellites that orbited the planet, we equipped them with radios linked to hyperaccurate atomic clocks. Today these satellites broadcast their signals to cheap electronic devices carried by soldiers and even schoolchildren, slowly making navigation an art of the past.

The mastery of radio began with nineteenth-century scientists, as they began to understand the nature of electricity and its twin brother, magnetism. In fact, serious study of these subjects commenced in Europe in the 1600s, but true understanding and practical application did not take hold until two centuries later.

An important step in the study of electricity occurred in 1800, when Italian scientist Alessandro Volta invented the storage battery. Instead of cranking crude generators to build up brief jolts of static electricity, researchers could simply attach wires to a set of Volta's batteries, which used zinc and copper plates dipped in saltwater to create a constant stream of electrical energy.

Two decades later Dutch scientist Hans Christian Ørsted noticed that when he connected a battery to an electric circuit, the needle of a nearby magnetic compass would shift its position. When Ørsted disconnected the circuit, the compass needle again pointed toward magnetic north. He had accidentally discovered that electrical currents generate a magnetic field.

Eleven years later English physicist Michael Faraday made essentially the same breakthrough from the opposite direction. Faraday found that by moving a magnet inside a coiled piece of wire, he could make an electric current flow through the wire. He quickly published his results and today gets credit for inventing the forerunner of today's electrical generators.

Between them, Ørsted and Faraday had established that electricity and magnetism were connected. It took a still greater scientist, Scottish physicist James Clerk Maxwell, to understand and explain the connection. Beginning in the 1860s, Maxwell presented mathematical proofs that electricity, magnetism, and even light were related to one another and were part of a family of physical forces that propagate through space at the speed of light, forces we now call the electromagnetic spectrum. Apart from the electromagnetic waves that were familiar to him, Maxwell's equations predicted that others were yet to be found.

For decades to come, scientists filled the gaps in Maxwell's spectrum, discovering hitherto unsuspected electromagnetic forces—infrared and ultraviolet light, X-rays and gamma rays. However, the discovery of greatest benefit to navigators came from a German physicist named Heinrich Hertz, who came up with a new way to test Maxwell's equations. In 1887 he set up an electrical apparatus that fired a spark across a gap between two wires. When Hertz switched it on, a similar spark appeared in a detection device he had placed on the other side of the room. Hertz had created electromagnetic waves and beamed them through the air to a receiving device a few feet away. His subsequent calculations found that the waves traveled at roughly the speed of light.

Hertz regarded his experiment as revealing little of practical significance, never exploring the real-world benefits of his discovery. He died in 1894 at the young age of thirty-six from an infection that had set in following surgery on an abscess in his jaw. News of Hertz's death stunned the scientific community, and the obituaries and encomiums flowed freely, eventually drawing the attention of a youthful Italian inventor who until then had taken little note of Hertz's work.

Guglielmo Marconi read of Hertz's spark experiment and quickly grasped the conclusion that had eluded the German scientist. If a spark could be transmitted across a room without wires, surely one could transmit telegraph messages in the same manner. It might be possible to refine the process, thus enabling the transmission of such messages over far longer distances—miles even.

Saying that Marconi invented radio is like saying that Columbus discovered America. The New World was already well populated by the time the great Italian navigator arrived. Likewise, plenty of people before Marconi had tried their hand at developing a means of wireless communication.

Marconi's achievement was the construction of radio gear that could send messages over long distances. A hands-on inventor with little formal scientific training, Marconi was a relentless experimenter who tried every possible method to extend the range of his transmitter. His breakthrough came in 1895, when by sheer chance he attached his antennae to the earth, forming a grounded electrical circuit. The range of the transmitter dramatically increased, and Marconi was able to send telegraph messages for miles.

The Italian government and business community were unimpressed by Marconi's work, but the British saw things differently. Marconi soon won the support of William Preece, chief electrical engineer of the British Post Office. Telegraph and telephone services were private-enterprise operations in the United States from the beginning, but on the other side of the Atlantic, privatization did not occur until the 1980s. Until then telecom networks in Britain were a government monopoly managed by the same organization that delivered the mail, and Preece was the post office's top technical man. Therefore, when Preece began showing Marconi and his inventions off at public lectures in London, scientific elites in Europe and the United States took note.

For years thereafter, Marconi and his colleagues steadily improved the range and transmission speed of his wireless telegraph system. When he first managed to transmit from Britain to North America in 1903, signal quality was lousy and breakdowns were frequent. It was only in 1907 that Marconi set up the first commercial radio communications service that spanned the Atlantic Ocean, from Ireland to Nova Scotia in Canada—a distance of nearly two thousand nautical miles. Two years later Marconi was awarded the Nobel Prize in Physics.

The journey from Hertz's primitive spark-gap experiments to transatlantic telegraph service had taken nearly twenty years. But now the pace of innovation accelerated. Researchers were quick to realize that radio would make an excellent navigational tool. Radio broadcasts traveled in a straight line from a fixed point. If one knew the location of the transmitting antenna, one could take aim at the signal and follow it home.

In 1902 American engineer and inventor John Stone earned the first US patent for a radio direction-finding device.[2] By 1906 he had persuaded the US Navy to give it a try. The device was installed on the USS *Lebanon*, an unglamorous vessel that had spent the Spanish-American War delivering coal to warships and later served as an all-around cargo carrier.

The Stone direction finder fell far short of expectations. His method required an array of two or more antennae that had to be aimed at the source. As the antennae rotated, the difference in signal they received would be used to focus on the source of the broadcast. However, in a design misstep, Stone's direction-finding antennae were firmly fixed to the *Lebanon*. They could be aimed only by rotating the entire ship. It was as if a ship's navigator could get an accurate compass reading only by turning his vessel's prow to the north. In addition, magnetic fields stirred up by other onboard electronics caused significant errors, even when the *Lebanon* was correctly lined up. Stone was destined for a long and distinguished career in telecommunications, but his direction finder proved a failure.[3]

A host of other inventors in the United States and Europe proposed their own radio direction finders. The American engineer Lee de Forest, whose vacuum-tube designs spawned the consumer electronics industry, patented his own direction system in 1904. Yet it would ultimately take a couple of Italians working in Paris to create a flexible system that could reliably detect a signal. In 1907 engineers Ettore Bellini and Alessandro Tosi developed a direction-finding receiver that combined a pair of fixed antennae with a charged coil that could be rotated to focus electronically on the incoming radio signal. This coil was attached to a compass-like pointer surrounded by a 360-degree scale. The operator rotated the coil until the incoming signal reached maximum strength. At that setting the pointer was aimed at the radio source.[4]

With the Bellini-Tosi system there was no need to move the entire ship, or even rotate a heavy antenna array. This made it far easier to install and operate. Marconi, whose engineers had been working on directional systems of their own, immediately recognized the superiority of the Bellini-Tosi design. In 1912 he purchased the relevant patents and put his engineers to work on improving the system.

Among those assigned to the task was Englishman H. J. Round, a graduate of Imperial College London who had joined the Marconi Company in 1902. In 1907 he discovered that certain silicon compounds would glow when electricity was passed through them. Though nobody noticed at the time, Round had come across the basic principle that drives the light-emitting diodes now used in nearly everything electronic, from smartphones to TV sets.

At the outbreak of World War I in July 1914, Round was working on wireless voice communications. He told the *New York Times* in October of that year that Marconi had developed a system that would carry voices anywhere from forty to sixty miles. The Italian Navy had agreed to buy the system, so that groups of warships could communicate by voice even in the midst of battle—a radical innovation at that time. Round added that Marconi was ready to demonstrate a wireless telephone service capable of transmitting between Britain and the United States, but that wartime censorship regulations forbade the experiment.[5] Round soon joined the British armed forces and was assigned to military intelligence. His technical skills made him a treasured asset, for at the time everyone was racing to develop better military applications for radio location.

This was the first war in which aircraft played a major role, and it soon became obvious that accurate navigation was as difficult as it was essential. How could reconnaissance or bombing planes find their targets and return safely home, day or night, fair weather or foul? The obvious answer—fly low and follow highways, railroad tracks, or rivers—was hardly a solution at all. Railroad tracks with their polished rails could easily be seen from the air, even at night, making them excellent navigational aids. But it also made the tracks fine places to set up antiaircraft guns and wait for unwary prey. Enemy combatants fired at low-flying planes, and the flimsy aircraft of that time could be brought down by a well-aimed rifle bullet. Over the western front, altitude was life. Flying through clouds or at night meant a greater chance of coming back alive; moreover, the ability to mount operations around the clock and in bad weather meant you could hit the enemy more often and do more damage. All sides were searching for better aerial navigation systems.

However, navigation aids were slow in coming. Something as basic as a good map was hard to come by. In the prewar years pilots had found that traditional maps, created for use on land, did not work so well in the open cockpits of airplanes. It wasn't only that the maps tended to tear or blow away. Pilots wanted maps marked with earthbound objects that could be easily spotted from the air, like church steeples or large factories. They also needed warnings about hazards to aerial navigation, like electric power lines. Britain and France made considerable progress in developing aviation maps prior to 1914, but other nations, including Germany, lagged far behind.[6]

Pilots, like sailors, could navigate partly by keeping track of their speed and direction of flight and doing a bit of "dead reckoning." Accurate instruments for measuring an airplane's speed had not been invented yet, and wouldn't be until the 1930s. And magnetic compasses proved unreliable. The plane's metallic components, such as engines or bombs, would distort the local magnetic field. It was possible to compensate for this effect, just as shipbuilders did when they started building their vessels out of iron rather than wood. However, the magnetic compasses of the time had been designed for use on ships, where the compass would remain reasonably horizontal relative to the earth's magnetic field. Airplanes turn by banking to the left or right, causing the compass to move out of horizontal. As a result, standard seagoing compasses tended to twitch and jerk erratically during turns, making it tough for pilots to find the correct heading. For the rest of the war, engineers on both sides developed aircraft compasses to address the problem, though none was a perfect solution.

Given these variables, radio guidance seemed an attractive prospect. A network of Bellini-Tosi receivers could listen in on a pilot's transmissions, calculate his position, and radio the information back to him. Yet the radios of the day weighed seventy pounds or more, at a time when primitive aircraft could barely lift their human occupants and ammunition for their guns. Full-fledged radio direction-finding systems for small short-range aircraft were out of the question, forcing pilots of these craft to rely on maps, shaky compasses, and glimpses of the earth below.

But it was not long before radio navigation was adopted for larger aircraft—thus creating history's first aerial bombers. Raining high explosives from the air would become a grisly hallmark of twentieth-century warfare. The first such attack of World War I, on January 19, 1915, must have carried a special sort of horror. The aircraft that appeared in the sky over England's eastern coast were not the sleek, fast-moving planes of later wars; such aircraft did not yet exist. Instead, the Germans sent zeppelins—battleship-size monsters, held aloft by lighter-than-air hydrogen gas. The January 19 raid killed four people and terrified millions more. Although of limited military value, the zeppelin strikes came as a psychological shock. Mighty as the British Empire was, it could not secure its homeland from air attack.

The Germans planned to continue and even intensify their attacks. However, their efforts were frustrated by the problem of delivering the zeppelins to their targets. After traveling hundreds of miles, much of it over open water, the pilots were often hard-pressed to figure out where they were.

Consider that first raid, launched from Hamburg. The two airships soared out over the North Sea, then turned west, hugging the Dutch coast. For this leg of the trip, their navigators needed only a good map and sharp eyes. Then came about 150 miles of blank saltwater before they could reach the southeastern coast of Britain. Here the commanders, like sailors of old, relied on compass headings and dead reckoning to estimate their position. Hampered by the lack of a reliable airspeed indicator, the zeppelins' speed had been calculated back in Germany, by flying them a precisely measured distance over land. Thus, the captains knew that they could travel at forty-five to sixty knots in still air.

But the air is rarely still. Even mild gusts of wind could substantially alter the speed or course of these hulking gasbags. And though the world's weather services track wind speed and direction in great detail today, they didn't in 1915. Pilots would instead dip low over the North Sea and try to gauge the wind by studying the size and shape of the waves below.

What resulted, then, was more of a guessing game than a precisely executed air attack. The first two raiders were aimed at the seaports of the Humber estuary in northeastern England. One of the captains, noting from the wave crests that he was being blown farther to the south, settled for an attack on the coastal town of Great Yarmouth. The skipper of the second zeppelin got lost and ended up bombing various targets of opportunity—basically, villages where unsuspecting locals had left the lights on. The commander himself did not know what he had hit until days later, when he read British press reports about the raid.[7]

Clearly, aerial bombardment would amount to little more than random terrorism until pilots were able to locate, identify, and attack specific targets from the air. The technology of World War I was not nearly good enough; indeed, it would be another sixty years before true precision bombing was possible. In the meantime, the German zeppelin raiders did their best, using time-tested methods and radio.

In the spring of 1915, the Germans erected one radio station on the island of Borkum and another in the town of Nordholz, about seventy miles to the east. Now a navigator could use radio direction gear to get a fix on his location. At first, the zeppelin navigators did this by broadcasting location requests to the land-based stations. Each station would take a bearing on the source of the broadcast and use triangulation to calculate the zeppelin's position. The ground stations would then relay the information to the zeppelin. The Germans encoded the transmissions to conceal their contents from the enemy. Yet the very fact of the transmissions made it possible for British radio networks to calculate the zeppelins' position, by using the same triangulation method as the Germans.[8]

Later in the war the Germans would adopt a more secure system of ground-based transmitters that broadcast a nondirectional signal that traveled in all directions. They also used a ring of directional antennae to send radio pulses in a sequence that corresponded to the points of the compass. On the zeppelin the navigator carried a specially calibrated watch that displayed the points of the compass rather than the time. This watch was synchronized with the directional pulses from the radio station.

The navigator would start the watch when he heard the first nondirectional radio signal and stop it on hearing the second radio pulse, the one that came from a particular direction. The face of the special watch would show the bearing of the antenna that had sent the directional signal. By repeating the process with the other station, he would get two lines of bearing. The point where the lines crossed would be the location of his zeppelin.[9] The new method was more secure because it was passive; the zeppelin did not have to give away its position by sending out a broadcast.

Still, radio navigation frequently failed its users. Reception was sometimes spotty, and a navigator could easily mishear the radio tones and put the airship on a false bearing. And the radio bearings became increasingly less accurate as the airship got farther away from the broadcast stations in Germany.

There were other problems. Antiaircraft gunnery improved, forcing the zeppelins to fly ever higher—they flew at altitudes up to twenty thousand feet at times. There they would lose sight of the ground and its easily recognizable landmarks. The thin air at such altitudes is always bitterly cold; the liquid used to dampen the motion of the zeppelins' navigational compasses would sometimes freeze and render them useless, while crew members would be dazed and stupefied from sheer lack of oxygen. And, of course, there was the constant threat of dirty weather, a hazard made worse by the lack of reliable weather forecasts. On October 19, 1917, an eleven-ship zeppelin raid fell prey to unexpected high winds over the North Sea. Most were blown off-course; four crashed or were shot down after straying into France. A fifth crash-landed in Germany, but fully two hundred miles from where it was supposed to be.

While the airships blundered through the air, their intended victims often had a better idea of their location than their pilots did, thanks to the efforts of H. J. Round, now a captain in the British army's corps of engineers. He had been tasked with setting up a network of radio stations in France to spy on the Germans. Incoming radio traffic was copied down and forwarded to "Room 40," the Royal Navy's cryptographic center, which did a masterful job of cracking enemy codes throughout the war.

Even when messages could not be read, Round's network proved immensely valuable, because each of his stations used Bellini-Tosi radio direction finders. While a single Bellini-Tosi rig could reveal the direction from which a radio signal came, two or more such systems at separate sites could calculate the broadcaster's location. Just draw a line on a map along the bearings taken at each receiving station; the transmitter would be located at the point where the lines crossed.

This network of Bellini-Tosi stations enabled the British to draw radio maps of German troop deployments and airfields and update them weekly. They were even able to identify the locations of zeppelins in flight and warn citizens of incoming raids. The system worked so well that the Royal Navy ordered Round to set up a similar network along the east coast of England to pick up German radio traffic coming in over the North Sea. Here the direction-finding gear proved its worth in spectacular fashion.

The British hoped to starve Germany into defeat by using its Royal Navy, the world's mightiest at the time, to enforce a blockade. In the years prior to the war, the Germans had built up a powerful fleet of its own, but were reluctant to take on the British in a full-scale battle. Instead, they relied on submarines—U-boats—to enforce an effective blockade of their own, sinking hundreds of ships en route to deliver food, fuel, and weapons to Great Britain. In addition, the Germans hoped to lure part of the British fleet into the North Sea and destroy it through combined submarine and surface attacks.

On May 30, 1916, the Royal Navy's network picked up a spike in radio traffic from the German port of Wilhelmshaven. Admiral Sir Henry Jackson, Britain's first sea lord, was troubled by the news. "The time was a critical and anxious one in the war," Jackson wrote in 1920, "and I had also some reasons for expecting that the German fleet might put out to sea during the week."[10] Messages intercepted by Room 40 indicated that eighteen German U-boats had put to sea a couple of weeks earlier, but had not shown up on normal trade routes for their usual attacks on merchant ships. The Royal Navy suspected that the U-boats were being deployed to assist the German fleet in their planned breakout.[11]

Later that day Round's radio listening posts issued a new report. The ship that had broadcast the previous message had since moved a few miles north. "Evidently she and her consorts had left the basins at Wilhelmshaven and had taken up a position in the Jade River, ready to put to sea," Jackson wrote. He decided to send the Grand Fleet to sea to try to meet the German fleet and bring it to action.

The next day the British fleet was on the move. Yet, incredibly, a later signal to the fleet's commander, Admiral John Jellicoe, mistakenly asserted that the German fleet was still in port. When Jellicoe came across the Germans at sea, three hours later, his confidence in radio intercepts was shattered.

The fleets engaged off the Danish peninsula of Jutland, in the largest sea battle of the war, and one of the largest in history. After a brutal but inconclusive clash, Room 40 intercepts revealed that the German fleet was preparing to dash back to the safety of Wilhelmshaven. The messages even included the Germans' intended course and speed. Jellicoe might have intercepted them and forced a decisive showdown. But given the earlier incident, he refused to believe the intelligence data, which in this case was absolutely correct.[12]

On paper Britain lost the showdown at Jutland. The Germans sank fourteen of the Royal Navy's ships at the cost of eleven of their own. The British lost more than six thousand sailors compared to only twenty-five hundred German dead. And the German fleet survived essentially intact. Yet the British achieved their most important objectives. The Battle of Jutland kept the German surface fleet in check and ensured that the British blockade remained as strong as ever. The result was economic collapse and mass starvation in Germany, a major reason for the country's ultimate defeat.

By the war's end, all combatants had made great progress in developing radio navigation and location gear. However, the resulting systems—while they sufficed for wartime—were not accurate enough to meet the needs of the emerging industry of commercial aviation. The war had spawned aircraft factories that could manufacture thousands of planes a year; these companies wanted to stay in business by building planes for hauling freight and passengers. In addition, thousands of aviation veterans wanted to keep flying. There were potential customers aplenty, if flying could be made a safe and reliable means of travel. What the airlines needed was somebody willing to go first—a paying customer with deep pockets and a willingness to back a daring and dangerous new technology. And they found it in the US Postal Service.

In the nineteenth and early twentieth centuries, when the postal service was known as the US Post Office Department, this federal

agency was crucial in building the nation's transportation infrastructure. Article I of the US Constitution authorizes Congress "to establish Post Offices and post Roads"—not only the infrastructure for
collecting and sorting the mail, but also the roads needed to move it.
From the beginning the post office played a major role in laying out
the nation's road network. Later, it indirectly propped up the infant
railroad industry by shipping the mail by train. Next came direct
subsidies to American shipbuilders to make their vessels competitive with Europe's best.[13] And less than a decade after the Wright
brothers got airborne, Postmaster General Frank Hitchcock began to
agitate for the establishment of a new service to deliver the mail by
air. On September 23, 1911, in Garden City Estates, Long Island, New
York, Hitchcock handed a sack of mail to pilot Earle Ovington, who
promptly lifted off and flew five miles to the town of Mineola, where
he tossed the bag from his cockpit to the ground below. It was the first
official delivery of US airmail.[14]

However, moving a sack of mail five miles by air was as wasteful as
it was trivial. A car or a fast horse would serve as well over so slight
a distance. Airmail was economical only for longer trips, such as the
two-hundred-mile run between New York and Washington, the nation's first regular airmail service, launched in May 1918. Longer trips
would be even better—New York to Chicago, for instance, a distance
of about seven hundred miles.

The trouble lay in finding one's way. In daylight, in good weather,
and with nobody shooting at the plane, these were challenging but
manageable journeys. The pilot who proposed to make the voyage at
night or on a cloudy day put himself at considerable risk. Yet time
and again, pilots made the effort, goaded by pride, overconfidence, or
the commands of superiors like second assistant postmaster general
Otto Praeger, who would fire men who refused to fly. It is no surprise,
then, that of the first forty men hired to fly airmail in 1918, only nine
were still alive by 1925.[15]

Praeger's ruthlessness led to a high casualty rate, but he had a point.
For airmail to pay, the planes had to fly, no matter what. Railroad

trains, the primary method for moving mail between cities in those days, would keep going at night and in almost any weather. When planes were grounded by weather or darkness, moving mail by air was no faster than relying on rails. The US Post Office launched an airmail service between New York and Chicago in September 1919, but with the planes so often unable to fly, it was hard to justify the extra cost—eight cents per ounce, four times the standard rate for a first-class letter delivered by rail.

In 1925 a group of Chicago bankers explained the problem to Praeger's successor, Paul Henderson. As matters then stood, mail from New York arrived after the banks were closed, whether the letters moved by train or plane. If an airmail service could fly checks overnight between Chicago and New York for delivery the following morning, those checks would clear a day sooner, giving banks quicker access to the money. For that the bankers would gladly pay the extra postage.[16]

Henderson was already at work on a solution—an illuminated highway in the sky known as the lighted airway system. In 1923 the post office had begun to erect a series of fifty-foot towers, about three miles apart, over the nine hundred miles between Chicago and Cheyenne, Wyoming, a ten-hour flight for the aircraft of that era. Each tower was adorned with a brilliant acetylene gas light that revolved six times a minute. Five large landing fields were built along the route and thirty emergency strips, all of them brightly lit.[17]

By the following year, the Chicago–Cheyenne run could be made in the dark. Also in 1924 a Chicago–Cleveland route was added, and in 1925, not long after Henderson's meeting with the bankers, a lighted Chicago–New York airway was completed. The flight between America's two largest cities could be made in a single night; mail could travel from coast to coast in about a day and a half, compared to about four days by rail, and the US Post Office's airmail service began to make a profit.

Not long thereafter, Congress passed the Kelly Act, which required the post office to hand off its airmail operation to private companies working under contract.[18] This piece of legislation marked the birth of the airline industry in the United States. Companies like Varney Air Lines, Robertson Aircraft, Pitcairn Aviation, Colonial Airlines, National Air Transport, and Western Air Express were formed to move the mail. Over the years, through various mergers and acquisitions, these carriers would become known as United Airlines, American Airlines, Eastern Air Lines, Trans World Airlines, and Delta Air Lines.

Congress had decided that flying cargo, and eventually people, should be a free-enterprise operation. It also concluded that overall management of air traffic must be a federal responsibility. The 1926 Air Commerce Act required that the federal government lay out a network of air routes connecting the nation's cities and build and operate the system of navigational aids to help pilots find their way.

It soon became obvious that lighted airway beacons were not enough. At times of snow, heavy rain, or thick cloud, they were virtually invisible. One pilot supposedly joked, "If you can see lights, it shows you don't want them; if you can't it shows you do."[19] World War I had shown that radio beacons might offer a solution, even though the performance of such beacons during the war had fallen far short of their potential. In Europe the direction-finding stations had been turned over to civilian use. Aircraft of the time were still relatively small, with limited carrying capacity. There was still no room for the heavy direction-finding equipment needed to let a pilot lock onto ground-based radio transmitters. Instead, early aircraft used smaller, lighter devices that could simply transmit and receive radio messages. A pilot in need of a location fix would transmit his request. Two or more ground stations would pick up the transmission and calculate the bearing from the receiving station to the airplane. With the bearings from multiple receivers, the technicians at the ground stations could calculate the location of the aircraft. They would then radio the information to the pilot. By 1925 the British, French, Belgians, and Dutch had covered their countries with such networks; the systems were integrated, so that pilots could fly with ease across national borders.

In 1926 US secretary of commerce Herbert Hoover began pressing for the construction of radio navigational aids to replace the acetylene beacons. The US Army had such a radio system under development. Inspired by research first conducted in Germany in 1907, the army's system would use a set of antennae that would broadcast directional radio signals at right angles to one another. For instance, one set might broadcast directly north and south, so the other would beam its signals to the east and west. These transmitters would have a range of about two hundred miles.

Each transmitter would send a single letter, over and over, in Morse code. One would transmit the letter A, which in Morse is a dot and a dash. The other would send the letter N—a dash and a dot. The two stations would synchronize their transmissions, so that each would transmit during the momentary pauses of the other. As a result, if an airplane's radio picked up the signals of both transmitters at equal strength, the pilot would hear a steady tone as the Morse letters merged. If instead the pilot heard one louder than the other, he would know he was not perfectly aligned with the radio station and would turn left or right as needed to smooth out the two signals. The pilot would also have to pay attention to whether the merged signal grew stronger or weaker. If it weakened, the plane was heading directly away from the radio beacon. If the signal intensified, the plane was moving straight toward it. The pilot was "on the beam," an expression that quickly entered the language of a nation besotted with the marvels of radio and air travel.

Such a beacon could guide planes in any of four different directions, and thus came to be called the "four-course radio range" system. Unlike the lighted beacons, it could be used day and night, in any weather. There was no need to signal to someone on the ground for a location fix. Instead, a pilot whose plane was equipped with relatively cheap, lightweight radios could find his own way.

The aviation entrepreneurs badly needed more reliable navigation aids, so they welcomed a plan that required the federal government to build radio highways in the sky. As commerce secretary and later as president, Herbert Hoover championed an aggressive role for the government in managing the nation's aviation system. Private companies would ply the skies, but Washington would set safety standards and build a network of national air routes. The airlines need pay only for compatible radio equipment in each plane, while the feds picked up the tab for dozens of costly radio ground stations—for instance, the twenty-one that spanned the distance from San Francisco to New York in 1931.

In the 1930s the primitive radio navigational system was sharply improved with the development of instruments that eliminated the need for a pilot to listen to the radio signal. Now she could simply glance at a gauge where a pair of needles indicated whether the aircraft was on or off the beam. Although much better than earlier navigational tools, the system was far from perfect. The low-frequency radio signals were susceptible to static and distortion. Flying through mountainous terrain was especially difficult, as the radio waves would bounce off the slopes at odd angles. A pilot might suppose he was steering toward a radio beacon only to slam into a sheer wall of rock. Five airliners flew into mountains in a one-month period between December 1936 and January 1937.[20] Still, the four-course system, supplemented with compass, chart, and eyeball, was a massive achievement. It became the primary US air-navigation system in the years before World War II.

Early radio navigational aids involved the interception of radio waves beamed directly from a transmitter. However, some engineers foresaw that even greater benefits might be achieved by pursuing an indirect approach. They knew that radio was part of the same electromagnetic spectrum as light, which illuminates the world by bouncing off physical objects and streaming into our eyes. Might not radio work the same way? Surely, if you beamed radio waves at, for example, the side of a building, some of that radiation would bounce off. And just as our eyes can see objects by the light they reflect, the right kind of radio receiver could "see" them by capturing the reflected radio waves.

The eccentric inventor Nikola Tesla had grasped the idea almost at once. Writing for *Century* magazine in 1900, Tesla wrote that with radio, "we may produce at will, from a sending-station, an electrical effect in any particular region of the globe; we may determine the relative position or course of a moving object, such as a vessel at sea, the distance traversed by the same, or its speed."[21] Nearly two decades later, at the height of World War I, Tesla proposed that such a radio detection system could be used to hunt down and sink the German U-boats that sank so many ships on their way to Great Britain.[22]

Tesla merely speculated about a radio detector. It took a German inventor, Christian Hülsmeyer, to actually construct one. In 1904, on the banks of the Rhine River in Cologne, Hülsmeyer demonstrated the "Telemobiloscope," a device that probably counts as the world's first radio detector. The Telemobiloscope sent out radio waves that echoed off ships moving up and down the river. In addition, it included a receiver that picked up the incoming echoes and set off an electric bell whenever another ship was in range. It was a crude device, but quite good enough to help a ship avoid collisions in a fog-shrouded harbor. To Hülsmeyer's astonishment, Germany's navy and its merchant marine shrugged off the idea, saying it was too costly to implement. Hülsmeyer eventually gave up the project; only after the value of radio detection had been vindicated during World War II did the aged inventor receive a measure of recognition.[23]

Eighteen years later, two US Navy researchers were testing high-frequency radio communication gear on the Potomac River when a wooden steamer passed by. When it came between the transmitter and the receiver on the opposite bank, the steamer briefly disrupted the radio signal, and Albert Hoyt Taylor and Leo C. Young had an idea: a similar radio rig could act as a sort of burglar alarm that could prevent an enemy vessel from sneaking into an American harbor at night or in fog.

Like the Germans, the US Navy did not embrace the scheme. Yet it continued to back radio research, and in the 1930s Taylor and Young's work was resurrected. By then scientists had developed a way of sending out radio signals in carefully tuned pulses. These radio signals, unlike earlier continuous signals, could be adjusted to allow more time between pulses for reflected radio waves to hit the receiver. That meant objects could be tracked with greater precision and at longer distances. In 1934 Taylor, Young, and a third researcher, Robert Morris Page, built a pulsed radio system at the US Naval Research Laboratory that could detect an airplane traveling overhead.

The Americans had plenty of competition. Scientists worldwide had realized the potential of radio detection; crude systems were under development in Germany, Italy, France, and the Soviet Union.[24] But the most significant breakthrough came from Great Britain. Shaken by the devastation wrought by the Great War's zeppelin raids, British military planners knew that any future war would feature better planes and bigger bombs. And when in 1933 the Germans elected as their chancellor the bellicose Adolf Hitler, it seemed a good idea to upgrade the nation's air defenses. In January 1935 the British Air Ministry formed a committee to investigate every possible means of fending off aerial attacks.

One of those tapped to work on the problem was Robert Watson-Watt, head of the radio department of Britain's National Physical Laboratory. First approached by the government about the possibility of shooting down planes with an electromagnetic "death ray," Watson-Watt quickly dismissed the idea as impractical. He felt there was a much better chance of building a radio device to detect attacking aircraft. After a test in February 1935 showed promising results, Watson-Watt was put in charge of a research project to build practical radio detection gear.

It would be a long, hard slog. For one thing, the radio gear needed to be as small and light as possible, so that they could be placed in airplanes, where they could be used to search for enemy planes and ships. Then there was the matter of wavelength—the distance between the crest of one wave and the next, measured in meters rather than feet. The radios of the 1930s transmitted waves that were often dozens of meters in length. However, these long waves were susceptible to atmospheric distortion and interference. Transmitting long waves requires large antennae, not well suited for mounting on airplanes. In addition, long waves did not return clear, sharp images when bounced off remote objects. If several planes were flying in close formation, the returning echo, displayed on a TV-like cathode-ray tube, resembled a big blob.

These problems could be solved by broadcasting radio waves at shorter wavelengths and higher frequencies—microwave radio. A microwave signal was less subject to atmospheric interference, harder for an enemy to jam, and easier to fit on a plane. And it served up much finer detail as well. An air-defense network equipped with microwave radars would be able to identify individual planes in an attacking swarm.

Over the next half decade, Watson-Watt and his team devoted all of their resources to solving the radar problem. They gradually developed transmitters capable of shorter and shorter wavelengths and built lighter, smaller radio sets that could be mounted in airplanes. In 1937 the British government began to erect Chain Home, a network of shore-based radio detection stations that would prove vital in the coming war. One month before the outbreak of World War II, they began installing airborne radars into fighter planes.

Still, the most valuable prize had so far eluded the British and everybody else—a practical, lightweight microwave radio detector. That changed on February 21, 1940, when John Randall and Henry Boot of the University of Birmingham successfully tested one of the most significant inventions of the century: a new kind of vacuum tube called a cavity magnetron. This tube generated ample power for radio transmission; more important, it produced radio waves, not by the meter, but by the centimeter—9.5 centimeters for the first device. And it would be relatively simple to mass-produce the cavity magnetrons, fitting them by the thousands in planes, ships, and antiaircraft guns.

Unfortunately, Great Britain's manufacturers were already strained to the limit making other implements of war. And the cavity magnetron did not work perfectly; it suffered from inconsistent power output. The British badly needed help in working out the remaining bugs and beginning mass production of the devices. So a team of British scientists sailed to Canada in September 1940 and then traveled to Washington, DC, on one of the most crucial secret missions of World War II. They had been dispatched by Sir Henry Tizard, an

Oxford-educated chemist who chaired the British government committee in charge of air defense. In their possession were the crown jewels of Britain's military technologies—documents describing new kinds of plastic explosives, gunsights, submarine detectors, early research on the atomic bomb, and an early form of jet engine. Finally, there was a cavity magnetron, one of a dozen then in existence.[25]

The Tizard Mission would negotiate a deal unlike any in history. Great Britain would share with the Americans an extraordinary trove of scientific knowledge. In exchange, the Yanks would put their best scientists to work on perfecting the technologies and then mass-produce the results and provide them in quantity to the British. At the time the United States was officially neutral, and most Americans vehemently opposed involvement in another European war. The administration of Franklin D. Roosevelt, however, was firmly on Britain's side and determined to do all it legally could to help them defeat Germany. Besides, the British bounty was extraordinary, especially the cavity magnetron. American engineers too had been hard at work on microwave radio detection gear and were making progress, but the cavity magnetron was years ahead of anything they had.

So the deal was done. Before the year was out, a new Radiation Laboratory had been erected at the Massachusetts Institute of Technology. It became the nation's center for radio detection research and development and the birthplace of ever more capable systems for use in every phase of the war. As the Americans took the lead in developing the technology, they also made the final decision on what it would be called. The British had dubbed it RDF, the same acronym they had used for World War I–era radio direction-finding gear, in the hope that enemy spies would not realize that the new system was something different. US Navy lieutenant commanders Samuel M. Tucker and F. R. Furth called it radar, for radio detection and ranging. In July 1943 the British adopted the same acronym.[26]

Radar technologies helped turn the tide of the war, enabling Allied ships, aircraft, and antiaircraft guns to spot attacking enemies well in advance and fire on them with remarkable accuracy. With millimeter radar sets, an object as small as the periscope of a German U-boat could be spotted in the dead of night, enabling airplanes and ships to smash the attacker before it could open fire. Yet radar was no panacea. Using it properly required intensive training. Ocean waves could easily be mistaken for a surfaced submarine; flocks of seabirds could appear to be attacking fighters.[27]

Indeed, during the Battle of the Atlantic, a different kind of radio detection was often even more effective. German submarines chasing Allied supply convoys would routinely transmit radio messages to their bases or to other U-boats. Allied warships equipped with the HF/DF system, sometimes called "huff-duff," could get a bearing on such signals, even if the submarine made a brief broadcast. A destroyer could then race toward the submarine and try for a kill; at the very least, the attack would likely scare off the U-boat. Meanwhile, the convoy could change course to evade further attacks.

HF/DF was an immensely valuable supplement to radar. It had much longer range, for one thing. Also, it was a passive detection system that emitted no radio waves that might alert an adversary. Over time, the Germans developed detectors that could warn them if a nearby Allied ship was scanning the sea with radar. These were useless against HF/DF, which did not transmit radio waves but merely listened for them.

The combination of lightweight microwave radars, HF/DF, long-range reconnaissance aircraft, and the successful cracking of encoded German naval radio messages strangled the life out of the German U-boat campaign. And as radars became even smaller and lighter, they enabled weapons of astonishing precision, such as the proximity-fused antiaircraft shells that ripped apart so many Japanese planes over the Philippine Sea.

As with the cavity magnetron, the concept of the proximity fuse was a British innovation. Fearful of German bombing raids and well aware of the country's limited ability to defend against them, the Royal Aircraft Establishment needed an innovative solution. Two of the most creative came from William Butement, a New Zealand–born scientist who in 1939 was serving as principal science officer at Britain's Air Defence Experimental Establishment. Butement suggested to his colleagues that it might be possible to track antiaircraft shells by radar as they rose toward their targets. When a shell got close enough to an enemy plane, a person on the ground could detonate it by remote control so that the plane would be ripped up by the resulting shrapnel cloud.

There is an implausible, Rube Goldberg feel to this idea, which went nowhere. But Butement's second idea, while ferociously difficult to implement, would work far more simply. He proposed building a tiny radar set inside the shell itself. This radar would send out a steady, constant signal and would include a receiver to pick up any radio reflections. This receiver would detect the presence of a nearby object, like an enemy plane, and blow up the shell. With help from colleagues, Butement laid out the basic circuit design for such a self-detonating shell.

Merle Tuve, a physicist at the Carnegie Institution who had played a major role in the development of radar, knew nothing of Butement's work in 1939. Yet he had already begun research on a proximity fuse for use in antiaircraft guns. Tuve considered two approaches—a photoelectric type that would be set off by the shadow of a passing plane or a fuse that relied on reflected radio waves.

In the premicrochip age, any such fuse would rely on glass vacuum tubes. The main issue was whether such tubes could withstand the shock of being fired from guns. One member of Tuve's team, Richard Roberts, attached an ordinary tube to a lead block, suspended it from the ceiling of a lab, and fired a bullet into the block. The tube was undamaged, even though Roberts calculated that it had experienced acceleration equal to five thousand times the force of gravity. Roberts took another tube, attached it to a hemisphere of lead, and dropped it off the roof of a building onto a steel plate. This time the tube was accelerated to twenty thousand times gravity and still worked fine.[28] Roberts's research suggested that a radar-guided shell might not be so hard to build after all. When Tuve and his team actually shot prototypes from guns, however, they consistently failed. Curiously, the glass tubes were undamaged, although the wire filaments inside the tubes were crushed by the acceleration.

Still, the potential benefits of the fuse compelled them to keep trying. In late 1940 the Tizard Mission delivered Butement's radio-based proximity-fuse design to the Americans. Convinced it could be made to work, the US government poured hundreds of millions of dollars into the project. By 1942 the project had outgrown the Carnegie Institution and was moved to Tuve's alma mater, Johns Hopkins University in Baltimore. Tuve became the first director of the school's new Applied Physics Laboratory (APL), which would make another crucial step forward in the science of location in the postwar years.

Tuve had some extraordinarily smart people at his command—James Van Allen, of course, and a Brooklyn-born mathematician named Milton Friedman, the future Nobel laureate who would go on to become one of the twentieth century's most influential economists. Later in his life, Friedman would recall that the statistical techniques he devised to analyze the performance of the prototype shells would later prove useful in analyzing income-distribution data.[29] Staffed with such talent, the APL team made steady progress. Van Allen solved the filament problem when he realized that what needed fixing were not the filaments themselves, but the metal supports that held them in place. He designed a different kind of support, shaped like a spring, in hopes that its flexibility would absorb the shock of firing. It worked.[30]

The US Navy was America's chief customer for antiaircraft ammunition, for use against Japanese planes in the Pacific. But the navy wanted the new shells only if they were proven reliable. The APL team had to demonstrate that at least half the shells in any given batch would perform as specified, a low standard of reliability until you remember that thousands of shells would be fired in a typical engagement.

In the fall of 1942, nearly a year after America's entry into the war, Tuve's team hit the fifty-fifty mark. Mass production of the new shells began, just four hundred a day at first, rising to seventy thousand a day by war's end. Van Allen and some of his colleagues were dispatched to ships throughout the US fleet to train gunnery officers in the use of the new shells. It took some persuading, as the veteran officers tended to distrust any untried weapons. Worse yet, by 1943, failure rates spiked for some of the oldest shells in the inventory. It turned out that the batteries powering the fuse's radar system had gone flat. Van Allen led a fleet-wide program to replace the batteries in a quarter-million deployed shells.

Yet by mid-1944, when Van Allen stood on the bridge of the USS *Washington,* the proximity shells had proven themselves accurate killers that dramatically improved the effectiveness of antiaircraft fire. They were considered so effective, in fact, that the US Combined Chiefs of Staff were desperate to prevent the Germans and Japanese from duplicating them. To keep the shells out of enemy hands, proximity-fused ammunition was at first used only at sea, mainly in the Pacific.

The proximity fuse was brought to the European theater only in 1944, after the Germans launched a new bombing campaign in June. This time the bombs were delivered by V-1 pilotless cruise missiles. Hundreds of them smashed into southern Britain, killing thousands of civilians at little cost to the Germans. British fighters shot down hundreds of the V-1s; as they had no human at the controls, they were relatively easy to attack. However, too many of them still slipped through. And if they were shot down over British soil, it did not matter. The falling wreckage would still cause terrible damage. The British shifted their antiaircraft guns to coastal areas, in a bid to shoot down the V-1s over water, but this new tactic proved ineffectual. Prime Minister Winston Churchill begged for help from President Roosevelt. In response, the United States delivered a one-two punch of high technology: the latest antiaircraft radar system to improve the aim of the British guns and a supply of proximity-fused ammunition.

Almost immediately, the tide turned. The new gun radars and the proximity shells began blowing dozens of V-1s from the sky, even as Allied forces in France systematically overran the German launching stations. In the last major land-based V-1 attack against Britain, the Germans launched 104 missiles; the guns shot down 68 of them. Fighters, barrage balloons, and malfunctions accounted for all but 4 of the rest. The proximity shells that missed their V-1 targets plunged harmlessly into the English Channel, posing no risk that the enemy might recover a few. Still, the military was reluctant to allow the use of the fuses over land. That finally changed in the autumn of 1944, when the Germans began bombarding the crucial port of Antwerp, Belgium, with V-1s and the antiaircraft gunners were issued proximity-fused shells to improve their kill rates.

The final barrier fell in December 1944, during the last great German offensive in the Ardennes. American forces were nearly overwhelmed in the early days of the attack, but responded with antipersonnel artillery shells equipped with proximity fuses. Instead of exploding when they hit the ground, these shells burst in the air above enemy troops, spraying them with deadly shrapnel.

Historians still debate whether the proximity shells were a major factor in the German defeat. General George Patton said, "The funny fuze won the Battle of the Bulge for us."[31] Other historians consider the shells helpful but far from decisive.[32] Tuve certainly believed in the deadly consequences of his work, and it troubled his conscience. "Our job was antiaircraft and then of course later the antipersonnel, that was a bit harder to swallow," he said in a 1967 interview. "I've never visited Germany. There are too many—too many orphans over there on my account."[33]

WORLD WAR II FINALLY ENDED FIFTY YEARS AFTER THE HAPPY ACCIDENT that helped Marconi transform radio from a scientific curiosity to an essential tool of communication and navigation. Yet in the early days of flight, when radio coverage was sporadic or nonexistent, scientists sought other methods of navigation, self-contained systems that would not require signals from a ground-based transmitter or an orbiting satellite. Something as simple as a rapidly spinning wheel could guide a traveler home.

Out for a Spin

On August 3, 1958, William Anderson knew exactly where he was. This was a remarkable feat, considering that he was moving at more than twenty miles per hour through one of the most hostile environments on the planet—the Arctic Ocean. And not on the surface. Captain Anderson and his 115 comrades aboard the submarine USS *Nautilus* were four hundred feet deep, swathed in near-total darkness, and locked beneath a vast, impenetrable sheet of ice. Still, at 11:15 p.m. eastern time on August 3, Anderson and his crew reached the North Pole, becoming the first ship in history to sail there. The question is: how did they know they had reached it?

It's no easy question. Many efforts to reach the pole have been fraught with controversy. For generations American schoolchildren learned that Robert Peary and his African American colleague Matthew Henson were the first to reach the North Pole on foot in 1909. But today many historians think they missed the mark, due to inaccurate navigation and faulty record keeping. Similar doubts have clouded the reputation of American aviator Richard Byrd, who in 1926 claimed the first airplane flight over the pole.

Byrd and Peary at least could rely on the sun, moon, and stars to guide them, but these aids were useless in a submerged submarine. For the same reason, the *Nautilus* could not count on guidance from radio beacons; besides, there weren't any in that part of the world. And no polar explorer could count on a magnetic compass; they become utterly unreliable as the voyager approaches the poles. Yet the crew of the world's first nuclear submarine traveled deaf and blind for five days and more than eighteen hundred undersea miles. When the *Nautilus* broke the surface again, off the coast of Greenland, it was a few miles from where it was supposed to be. Another nuclear sub, the USS *Skate*, repeated the voyage in March 1959. The *Skate* actually surfaced at the pole and confirmed its location through celestial navigation.

These historic feats of advanced navigation were made possible not by complicated electronics, but by the most primitive of machines— the wheel. One of mankind's earliest inventions, the wheel is at least six thousand years old and has been used since its creation to move travelers and cargo from place to place. However, it was not until the nineteenth century that a rotating wheel was recognized as being useful in navigation.

Before the navigational wheel came the pendulum. In 1851 French physicist Jean Foucault successfully demonstrated that the earth rotated on its north-south axis. By the mid-nineteenth century, everybody knew this to be true, but Foucault devised a way to watch it happen. Hanging a large pendulum from the ceiling of the Pantheon in Paris, he set it swinging back and forth. Anybody might expect the pendulum to pass over the same strip of flooring, again and again, but this commonsense expectation turned out to be wrong. As the hours passed, the pendulum's position relative to the floor began to change. As it swept back and forth, it shifted.

You can envision the pendulum swinging back and forth across the face of a clock. At first it might pass over the numbers 12 and 6, but it would slowly shift until it swept across the 1 and the 7, the 2 and the 8, the 3 and the 9, and so on, until it made a complete circuit.[1] If

Paris had been at the North Pole, latitude 90 degrees north, the cycle would have taken about twenty-four hours. When the pendulum is farther south, the process takes longer. In Paris latitude 48.87 degrees north, it took about thirty-two hours. On the equator, at 0 latitude, the pendulum would not shift at all. And in 2001, when some scientists from Sonoma State University hung a pendulum very close to latitude 90 degrees south—the South Pole—it made the circuit in twenty-four hours and fifty minutes, only counterclockwise.[2]

What Foucault's experiment showed was that the arc of the pendulum did not change its direction at all. Centuries earlier, Sir Isaac Newton had explained why, in his principle of inertia. Once set in motion, the pendulum would keep moving in the same direction, unless some force altered its course. No such force existed for Foucault's pendulum.

As the pendulum swung, the planet itself was shifting beneath it. At the North or South Pole, the shift would synchronize with the earth's rotation around its axis. At latitudes between the poles, the effect would be slightly offset by the fact that the building housing the pendulum would shift eastward as the earth rotated. Thus, if you set up a chain of pendulums from the North Pole to the equator, the pendulums farther south would take more time to complete their circuits until you reached the equator. There the pendulum and the earth would both be heading east at the same speed, and there would be no apparent shift in the pendulum's swing at all.

Foucault had come up with a clever way of making the earth's rotation visible. Still, he wanted a better one. The fact that the pendulum gave different results at different latitudes was confusing, and Foucault sought a device that would synchronize with the earth's rotation at any latitude.

What he found was a rapidly spinning globe developed decades earlier by a German scientist. Johann Gottlieb Friedrich von Bohnenberger, a math and astronomy professor at the University of Tübingen, had built a device in 1817 that featured a ball of ivory that could rotate on a central spindle. This sphere was mounted in a set of rings, or gimbals, that let the ivory ball move freely in all directions. Once the ball was set to spinning, it did not shift inside its gimbals. On the contrary, the axis about which the ball spun would keep pointing in the same direction, and only a firm shove could alter its position.[3] Foucault instantly grasped the significance of Bohnenberger's device, which the German scientist had simply called "the machine." Like Foucault's pendulum, the spinning disk was the captive of inertia and could also be used to detect planetary rotation.

Foucault's version of Bohnenberger's machine replaced the ivory ball with a wheel—a disk pierced at its center with a spindle—and retained the gimbals that let the disk and its spin axis point in any direction. Further, he added a microscope that enabled the user to measure any minute change in the direction the axis was pointing. Foucault lacked a way to keep the disk spinning rapidly for more than a few minutes, and he knew that in that time, any shift would be very slight.

Now it was a matter of giving the disk a good spin and watching the results. Sure enough, the microscope detected a small change in the orientation of the spin axis over time. Yet no force had acted on the disk to change its direction. In fact, it was pointing the same way all along. What had moved was everything else; thanks to the free-moving gimbals in which it was mounted, the wheel remained stable even as the planet moved around it.[4]

Foucault called the new machine a gyroscope, from the Greek words for rotation and vision. In effect, it was a "see-it spin" machine, designed to demonstrate at a glance the rotation of the earth. Unlike the pendulum, the gyroscope was unaffected by its latitude. It worked the same way anywhere on the planet.

A man as sharp as Foucault quickly realized that his gadget was good for more than a clever science experiment. For centuries sailors had been plagued by the unreliability of magnetic compasses. The problem worsened in the nineteenth century as shipbuilders abandoned wooden hulls for iron and steel, which scrambled the planet's magnetic field. There were ways to compensate, and sailors accordingly recalibrated ships' compasses. However, the results were far from perfect, and even a variation in the kind of cargo on board could alter compass performance. The gyroscope was a universal solution, a new kind of compass that always pointed in the right direction, regardless of the kind of metal in the ship's hull, its hold, or even in the pockets of its crew. A gyroscopic compass would always point in the same direction, no matter what.

Better yet, the gyroscopic compass could be modified by adding weights to the spin axis. Putting any kind of force on the gyroscope axis causes it to rotate 90 degrees from the direction of the force, in a process called precession. With carefully balanced weights, the gyroscope would precess until its axis pointed north. But a gyroscopic compass would point to "true north," the direction of the geographic North Pole, not the magnetic North Pole sought out by magnetic compasses. Apart from being in a different place from the geographic pole, the magnetic pole is constantly moving, due to shifts in the planet's magnetic field. And magnetic compass readings become less accurate as the navigator approaches the magnetic pole. A gyroscopic compass would remedy the two great weaknesses of its magnetic counterpart— it would always point to true north, and its accuracy would be unaffected by magnetic-field fluctuations.

Although he understood the value of his invention, Foucault never cashed in. Blame it on technology—in this case, the lack of it. A gyroscopic compass would have to spin nonstop throughout a long voyage, but in the 1850s nobody had figured out a way to keep it spinning. Within the decade scientists were designing electric motors to do the job. Later gyros were powered by a constant jet of compressed air. A hundred years later scientists would eliminate spinning wheels altogether and build gyros that relied on rotating laser beams. It all came too late for Foucault; he died in 1868. And it would be another forty years until the gyroscope was put to work as a navigational aid, by a most unusual character.

Born in 1872, Hermann Anschütz-Kaempfe studied art history and medicine in college, but gradually became obsessed with a field of endeavor as exotic as it was impractical—polar exploration. He was not alone. In the late nineteenth and early twentieth centuries, a host of ambitious men, including Fridtjof Nansen and Robert Peary, vied to be the first to reach the North Pole. Some tried to get there on foot, while others attempted to sail there in specially reinforced ships capable of smashing their way through the ice. In a paper published in 1902, Anschütz proposed a radical alternative. More than a half century before the voyage of the USS *Nautilus*, he argued that the pole was best approached from below, by sailing a submarine under the ice cap.[5]

Given the crude state of submarine technology at the time, it would have been a harrowing voyage. With its nuclear reactor, the *Nautilus* could remain submerged for months and travel hundreds of miles in a day. The crew traveled in comfort: the reactor provided ample electric power for heating and air-conditioning; a distillation system created a steady supply of fresh water for drinking, bathing, and cooking; and the crew had no trouble breathing, thanks to a nuclear-powered electrolysis unit that produced oxygen from seawater.

An early-twentieth-century submarine offered none of these amenities. Neither was it equipped, as the *Nautilus* was, with sonar equipment to let navigators identify undersea obstacles, or closed-circuit television cameras for inspecting the underside of the ice. Yet naval engineers at the Krupp shipbuilding plant assured Anschütz that they could build a submarine capable of making the polar run. It would be an uncomfortable and risky voyage, but it could be made.

However, when Anschütz proposed his plan at a meeting of the Vienna Geographic Society, a member of the audience called out, "That's all very fine, but how are you going to steer your submarine?"[6] Sealed under dozens of feet of ice, how would the submarine's officers know which way to sail or whether they had actually reached the pole?

Submarines frequently surfaced in those days, and not only to allow the crew to breathe fresh air and to recharge the batteries on the boat's diesel engine. They also bobbed up to the top so the boat's navigator could glance at the sun or stars or landmarks on a nearby shore and in that way figure out where they were and which way they were headed. The magnetic compass wasn't much help. As tricky as they were on surface ships, they were even less reliable on submarines, which were effectively large steel tubes tailor-made to screw with the planet's magnetic field.

Anschütz saw a solution in Foucault's work with gyroscopes. A gyroscopic compass could reliably steer its user due north. Anschütz was not the first to think of using a gyroscope to navigate a submarine. A battery-powered boat called the *Gymnote* joined the French Navy in 1888, the first submarine ever to put to sea on behalf of a major nation. Among its other innovations, the *Gymnote* used a gyroscope to steer when submerged. The performance of the device was less than satisfactory.[7] Anschütz was confident he could do better.

His confidence was justified. By 1903 the self-taught engineer was conducting preliminary tests on such a compass on a lake near Munich. By 1904 he was testing an early model on a German navy ship. And by 1905 Anschütz had launched a new company to market his invention. By then he had forgotten about trips to the North Pole. Anschütz realized that he had a decent chance of becoming very rich, as navies and merchant fleets embraced his invention, by then called the gyrocompass.

In 1908 a gyrocompass made by Anschütz & Company was installed aboard the battleship *Deutschland,* flagship of the German fleet. After a twenty-eight-day trial, the captain praised the performance of the new instrument, legitimizing it with navigators around the world. Building gyrocompasses was about to become a large, lucrative business, big enough to attract an American entrepreneur who already knew a thing or two about compasses.

Born in the central New York town of Cincinnatus in 1860, Elmer Sperry was an incorrigible entrepreneur and inventor. By the time of his death in 1930, Sperry had earned more than four hundred US patents in fields ranging from mining machinery to electric motors, from generators and lighting to industrial chemicals, and finally, of course, gyroscopes. Sperry first took note of gyroscopes in 1907, after buying several toy versions for his sons, Edward, Lawrence, and El-mer. This inspired him to read up on the latest European experiments with gyros, including an effort to use the devices to stabilize an automobile or ship.

Recall that when a force is applied to a spinning gyroscope, it tends to precess 90 degrees from the direction of the force, while the built-up momentum of the spinning disk pushes back against the force. In the early 1900s scientists in Germany and England realized that this physical property could become a blessing to the seasick traveler.

Imagine a ship with a big gyroscope mounted in its hold, its spin axis pointed straight up, so that the gyroscope disk is parallel to the deck. Now put the gyroscope in a gimbaled bracket that runs from left to right across the width of the ship. If a large wave hits one side of the boat, it will naturally tend to rock from side to side. When that force hits the spinning gyroscope, it will precess inside its gimbals, tilting as its built-up momentum pushes back against the roll. If the gyroscope disk is heavy enough, and spinning fast enough, its inertia will absorb much of the rolling force and stabilize the entire ship.

Sperry first planned to build gyrostabilizers for cars. In a 1907 US patent application for such a contraption,[8] he calculated that a two-hundred-pound gyro would produce enough momentum to push back against a roll with four tons of counterforce. Sperry secured his patent, but gyrostabilized cars went nowhere; four-wheeled vehicles with low centers of gravity tend to remain upright without the extra help.

Ships at sea were another matter. On commercial vessels, side-to-side rolling in heavy weather was a hazard to the crew and a misery to the passengers. For the world's navies, ship stabilization was a matter of national security. The early twentieth century saw the start of a massive naval arms race, with Britain, Germany, the United States, and Japan launching a host of heavily armored, heavily armed battle-ships. These mighty vessels wouldn't be nearly so impressive if their twelve-inch guns couldn't hit their targets. Accurate gunnery would be far easier on a ship that did not roll.

German engineer Ernst Otto Schlick thus had little difficulty in interesting his country's navy in gyrostabilization. At the same time, in 1908, Sperry targeted the American naval market with a rival product that was decidedly better than Schlick's. The Sperry gyro-stabilizer did not wait for the ship to start rolling; instead, it used a smaller sensitive gyroscope to signal the start of a roll. This would activate a motor that would shift the far bigger main gyroscope, caus-ing it to precess in advance of the roll. As a result, the Sperry gyro-stabilizer was more responsive and delivered a smoother ride.

While still perfecting his gyrostabilizer, Sperry turned his attention to the gyrocompass, a product with even greater commercial appeal. Sailors could learn to live with a rolling deck, but accurate navigation was essential. Sperry had met Anschütz that year, during a visit to Europe, and came away unimpressed with the German's skimpy un-derstanding of science and engineering. If this dabbler could build a halfway-decent gyrocompass, Sperry figured he could do far better.[9]

The US Navy had been on the verge of ordering Anschütz & Com-pany gyros until Sperry came along with an American alternative. In the spring of 1911, the navy was running sea trials on the Sperry compasses. They performed so well that Sperry received orders for six of them, which he began to deliver by year's end.

Sperry obtained patents in the United States and Great Britain for his gyrocompass and set about selling them to the world's major navies, including the Germans. In May 1914, a few months before the outbreak of war, the German Imperial Navy acquired a Sperry gyro-compass. To the navy it was probably just a prudent way of locking in a second supplier for vital technical equipment. The United States, af-ter all, was a neutral party at this time, happy to sell its technology to Britain and Germany alike. To Anschütz, however, the loss of a single sale to a non-German rival was an unbearable insult. Anschütz was certain that the Sperry product violated his own company's German and British patents. He filed suit against Sperry in both countries.

The suits dragged well beyond the start of World War I, with pre-dictable results. Sperry was victorious in Britain, but beaten in Ger-many. If local nationalism wasn't enough of an edge for Anschütz, he also benefited from the testimony of his expert witness, a veteran Swiss patent inspector named Albert Einstein.[10] Ultimately, it did not matter. With the British wartime blockade, Sperry couldn't sell in Germany. The British and Americans were eager to buy, and after the war the victorious Allies seized all German patents, ensuring that the judgment against Sperry was never enforced.

Sperry also pioneered the use of the same technology in airplanes. Once again, Sperry began with the idea of a gyrostabilizer, a machine that would enable aircraft to fly straight and level even when buffeted by unpredictable gusts of wind. His early experiments were unsuccessful, but in 1912, surrounded by aviation experts, Sperry gave it another chance. He had befriended Glenn Curtiss—the most revered American aviation entrepreneur at the time, next to the Wright brothers. Another friend of Curtiss, Washington Irving Chambers, was a US Navy captain who had been put in charge of research in naval aviation. Finally, there was Lawrence Sperry, Elmer's twenty-year-old son—handsome, impulsive, and a born aviator. In the summer of 1910, while his parents were away, Lawrence built the wings of a man-size biplane inside the family home. The glue he had used needed high heat to dry properly; Lawrence stoked the household furnace till it cracked. The wing couldn't fit through the doors of the house; Lawrence tore out a wall.

Enough was enough; his father exiled Lawrence to an Arizona boarding school, where he performed well but remained an aviation addict. The elder Sperry decided that if his son was determined to fly, he would be better off doing it under the eye of Glenn Curtiss. In the meantime, perhaps Lawrence could help his father design and test the much-needed aircraft stabilizer.

Between 1912 and 1914, the Sperrys developed the first autopilot system. At its heart were two sets of gyros, one that countered the plane's tendency to roll to the left or right, the other to prevent it from tilting up or down. Whenever the plane deviated from its proper course, the gyros, mounted in their gimbals, would keep pointing the same way, while the plane rotated around them. The shifting gimbals would trip an electrical switch that activated air-powered pumps to adjust the plane's control surfaces—the ailerons and rudder—and bring it back into alignment with the gyros.

In late 1913, in North Island, California, Lawrence decided the new machine was ready for a trial run. The test aircraft was an early "flying boat," capable of landing on water. Although Lawrence was convinced that the stabilizer would work, his chief test pilot, Patrick Bellinger, was not. Any hint of a deviation from course, and Bellinger grabbed the controls before the gyros could respond. One Sunday morning in November, Lawrence sneaked out to the plane and took off alone. Lying on the floor of the plane, where he could keep an eye on his gyroscopes, he nudged the stick and rudder to throw the plane off course. Time and again, it righted itself.

The men watching from the ground could see nothing of the pilot. It was as if the plane was flying itself. Indeed it was, as young Sperry demonstrated by walking onto its wing as it cruised above the water. The uneven weight distribution barely troubled the gyros, which stabilized the plane almost immediately. Lawrence conducted a similarly flamboyant demonstration in June 1914 at a major air show in Paris sponsored by the French War Department. The Sperry aircraft stabilizer won an award of fifty thousand francs; later that year it won the Robert J. Collier Trophy, the highest American honor for advancing the science of aviation.

The outbreak of World War I ensured demand for Sperry gyrocompasses and gyrostabilizers for warships, but the company was less successful at marketing its aircraft stabilizer. Pilots of the first primitive warplanes generally valued agility and maneuverability more than the ability to fly a steady course. The French military obtained forty of the aircraft stabilizers in 1916, but few of them were used in combat.[11]

The Sperrys, and the US military, took a longer view. They realized that self-steering aircraft might be developed into flying bombs— slow-motion artillery shells that could fly for hundreds of miles and then plunge to earth over an enemy city. There would be no need to risk the lives of airmen by sending them into hostile skies. A foe could be shelled into submission using unmanned planes, packed with explosives and steered by Sperry's gyroscopes. Lawrence Sperry patented the concept in the United States, and in 1917 Sperry Gyroscope won a US Navy contract to transform several seaplanes into "aerial torpedoes."

The development of such technology was tough going. It was not enough for the plane to fly straight and level; it must be catapulted into the air, like a modern jet from the deck of an aircraft carrier, but with nobody at the controls. After launch the plane must follow a preset course at an assigned altitude and travel a specific distance before shutting off its motor and plunging to earth. It was a difficult proposition, and British inventors had been doing research along similar lines in 1916 and 1917, only to give it up after repeated failures.

The Sperry-navy team did little better in their twelve launch attempts. However, they managed one crucial success. On March 16, 1918, one of the Sperry test planes lifted from its catapult, rose into the skies above Amityville, New York, and traveled the preset distance of one thousand yards before plunging into Great South Bay, precisely as intended.

Fifteen years after the first powered flight, the Sperrys had built and flown a completely automated aircraft, stabilized and steered by their gyros. More precisely—and more ominously—they had built the first guided missile.

Luckily for the combatants of World War I, the Sperrys and the navy were never able to build aerial torpedoes that could reliably travel hundreds of miles and hit their intended targets. The effort continued well beyond the end of the war, but by 1922 the navy shelved the program. A competing US Army program fared no better and was halted in 1920.[12] Nevertheless, the flying-bomb concept survived and eventually matured. Researchers in Britain, the United States, and Germany continued their experiments in the interwar years. The first operational flying bomb, the German V-1, was destined to vindicate the technology in horrific fashion. More than twelve thousand of these gyroscopically steered cruise missiles flung tons of high explosive into London and Antwerp between 1944 and 1945, killing thousands of civilians.

Despite its capacity for previously unimaginable destruction, the same gyroscopic systems became essential for safe and profitable commercial air travel. The directional gyro, unaffected by stray magnetic fields, became a vital supplement for the magnetic compass. And other gyro-based devices would protect pilots from one of their deadliest perils—their own senses.

People still speak of "flying by the seat of one's pants," no doubt forgetting that quite a few pilots have died that way. Many early aviators believed that they could accurately steer an aircraft by feel, even in darkness or bad weather. They relied on physical sensation to tell them whether the aircraft was rising or falling, turning left or right. Time after time, these bold, confident aviators were burned, crushed, or drowned, having learned too late that their senses had deceived them.

A US Army flight surgeon, David Myers, made it his job to figure out why. Myers suspected that there was something about flying that scrambled the human sense of equilibrium. Based on the movement of fluid in the inner ear, our sense of balance is tuned for life on land. Myers wondered if the movements of an airplane could deceive the balance system badly enough to leave even skilled pilots utterly disoriented.

In 1926 Myers began putting army pilots in a chair that could easily be rotated on a central axis, like the chairs at a barbershop. Then he had the pilots close their eyes while the chair was rotated. After a few spins, Myers would gently stop the chair's rotation without informing its occupant. Then he would ask the pilot whether the chair was still spinning. Time and again, the pilots insisted that it was, only to find themselves sitting still when they opened their eyes. Myers also found that by altering the speed of rotation, he could persuade many pilots that the chair had come to a halt while it was still spinning.[13] It became clear that after a little rough weather or poor visibility, many pilots no longer knew whether they were coming or going. Worse, they didn't know that they didn't know. They remained supremely confident in their skills, right to the fiery end.

The experiment convinced Myers and an army colleague, Captain William Ocker, that pilots needed instruments capable of confirming their aircraft's states of motion. Ocker tested his hypothesis by equipping a rotating chair with a black box in which the occupant would place his head, to ensure he could not see his surroundings. Inside the box he could see two simple instruments—a magnetic compass and a gyro-based device developed by Elmer Sperry called a turn-and-bank indicator. As the chair turned left or right, the turn-and-bank indicator's gimbaled gyro would precess, swiveling at a 90-degree angle to the turn. This precession would tug at a spring connected to an indicator dial, showing the user that his chair was turning, whether he felt it in the seat of his pants or not. Pilots who tried out the chair no longer had to guess whether they were turning. They could tell at a glance.

Ocker and Myers had proven that safe flying required pilots to ignore their untrustworthy senses and rely on instruments. Indeed, their research suggested that with sufficiently reliable instruments, a pilot need no longer rely on visual landmarks. Instead, he could use a combination of radio signals and onboard instruments to fly between two points in any weather, at any time.

First, however, they would need better instruments. The instability of the magnetic compass was still a problem. The turn-and-bank indicator was too limited a tool. Pilots wanted a device that would tell them a plane's orientation relative to the horizon. Just as they needed to know whether the plane's wings were banked left or right, they also wanted to know the angle of the plane's nose—too high, too low, or just right? In other words, they needed an "artificial horizon" device, rather like the one that a French admiral named Fleuriais had developed in 1885 for pilots of lighter-than-air balloons. A battery-powered air pump kept the gyro spinning; a line drawn on the side of the gyro formed the horizon, ensuring that balloon pilots knew the orientation of their craft relative to the ground below.

The Fleuriais gyroscopic horizon did not work very well. Yet by the late 1920s, Sperry engineers, led by Elmer Sperry Jr., had put together an artificial horizon so reliable that its descendants are still found in nearly every airplane. In addition, they had devised a hybrid solution to the instability of the magnetic compass. Ships at sea were able to solve the problem by relying on gyrocompasses. Airplanes posed a tougher problem. It took time for a gyroscope to realign itself to true north after a turn. The time lag was irrelevant when the gyrocompass was mounted on a slow-moving ship, but on an airplane, gyrocompass readings were inaccurate for some time after the aircraft turned. This delay made the gyrocompass no more trustworthy than the magnetic compass.

Sperry's engineers worked out a compromise solution. Aircraft would be fitted with magnetic compasses, carefully shielded and balanced to offset the effects of the plane's metallic components. These would reliably point to magnetic north when the plane was flying straight and level. Alongside the magnetic compass would be a simple directional gyro, which would be pointed at magnetic north with the aid of the magnetic compass. Mounted on gimbals, this gyro would keep pointing north regardless of the aircraft's motions. The directional gyro would be the primary instrument for measuring direction; the magnetic compass would be used to occasionally realign the gyro.

Between them, the directional gyro and the artificial horizon assured a pilot that he was traveling in the right direction and that his aircraft was flying straight and level. Combined with radio navigation aids, a reliable altimeter to measure height above the ground, and a device for measuring the plane's speed through the air, such tools would in theory allow a pilot to launch himself from the earth, fly for many miles, and safely land, all without seeing the world around him.

In 1929 blind flight became a reality. On September 24, US Army lieutenant James Doolittle climbed aboard a Continental NY-2 biplane at the army's Mitchel Field air base in Long Island, New York, with his copilot, Ben Kelsey. Kelsey rode along in the two-seater aircraft, but he never touched the controls. Doolittle, a world-famous aircraft racer who held a doctorate in aeronautics from the Massachusetts Institute of Technology (MIT), flew the plane on his own. The cockpit had been covered by an opaque hood. Doolittle intended to fly the plane with nothing to guide him but signals from a radio beacon and the gauges on his instrument panel.

And he did. In a flight that lasted fifteen minutes, Doolittle soared one thousand feet, circled the field, lined up on the runway, and made a landing that he himself graded as "sloppy."[14] No matter; the plane was right-side up, and its crew was alive. Doolittle became the first aviator to fly blind, guided entirely by electronic and mechanical instruments. It was probably the most important advance in aviation since the Wrights had left the ground. Yet Doolittle is barely remembered for this feat, because he outdid himself in spectacular fashion. Thirteen years after his blind flight, Doolittle led America's first bombing raid on Tokyo four months after the Japanese attack on Pearl Harbor. The feat earned him a Medal of Honor and elevated him to the pantheon of American military heroes.

The Doolittle Raid was a welcome propaganda victory for the United States, but had little military value. By contrast, nearly every airplane built since Doolittle's 1929 blind flight has profited from the lessons learned. Pilots still use the same basic suite of instruments to orient themselves and find their way safely back to earth.

These tools include the radio beacons discussed in Chapter 2, such as the four-course radio range system. Radios provided an external source of navigational data to supplement the information provided by cockpit instruments. Yet pilots began to wonder whether it might be possible to navigate without any sort of external aid. Couldn't engineers build a machine that would perfect the old sailing technique of dead reckoning? With precise knowledge of a craft's starting point, such a machine could calculate all changes in speed and direction after takeoff and thus the aircraft's location at any point in its journey without any external assistance.

At least one man had conceived of the possibility long before the invention of aircraft that could rise from the earth. In 1873 an Irishman named John Joseph Murphy sent a letter to the editors of *Nature*, then as now one of the world's leading scientific journals, responding to a letter written by Charles Darwin. The great biologist had noted that some animals, taken far from their homes, could make their way back, even though the animals had been transported in boxes that prevented them from seeing the path they were traveling. Darwin argued that therefore animals and humans must each possess some innate ability to home in on a destination without visual cues to guide them. Murphy retorted that while animals possessed this trait, humans did not. However, he added, it might be possible for humans to emulate the feat. "If a ball is freely suspended from the roof of a railway carriage," wrote Murphy, "it will receive a shock sufficient to move it, when the carriage is set in motion; and the magnitude and direction of the shock thus given to the ball will depend on the magnitude and direction of the force with which the carriage begins to move." Every motion of the train would disturb the hanging ball in a particular way, depending on the direction of movement, or the rate of acceleration. Now suppose a machine, connected to a highly accurate clock, could record the direction and force of each shock. It should be possible, Murphy concluded, to calculate from this data the exact position of the railway carriage at any point in its journey, with no need for external guidance. The train's direction might be read off a dial calibrated in degrees of a circle, while another dial would reveal the distance traveled in miles.[15]

Murphy was only trying to conceive of a mechanism that could match the remarkable navigating skills of animals. He had no intention of building one, realizing that "such delicacy of mechanism is not to be hoped for," certainly not with the components available in 1873. Yet his letter shows that the concept of a completely self-navigating machine was in the air long before Sperry or Doolittle.

Early in the twentieth century, at least five patents were filed in various countries covering self-contained navigation equipment. None of them was ever put into use, but all were based on the same basic concepts.[16] Start with a set of accelerometers, simple devices that measure the acceleration of the vehicle in which they are mounted. These accelerometers detect movements to the north, south, east, and west, as well as up and down. With each movement, the accelerometers would respond with an electrical signal proportionate to the force of acceleration.

Under some circumstances, these accelerometers would deliver misleading data. In an airplane taking off, for instance, an increase in altitude could be misinterpreted as forward acceleration, throwing off the accuracy of the system. To avoid this, the accelerometers must be mounted on a stable platform that would keep them level relative to the earth's surface. The platform would be mounted on free-swinging gimbals and equipped with a pair of gyroscopes. When spun up, the inertia built up by the gyros would stabilize the platform, ensuring that the accelerometers were constantly level. Now it would be possible to mathematically process the data from the accelerometers to calculate the vehicle's position at any moment.

In effect, the device would measure changes in the inertia state of the vehicle. Remember the old Newtonian law—objects at rest remain at rest until acted upon by an external force, and objects in motion continue moving in the same speed and direction until some force alters their movement. A self-contained navigation device would track the vehicle's location by detecting any changes in its inertia. Hence the name that was applied to the concept—inertial navigation.

A onetime actor and World War I pilot named Johann Maria Boykow took up the idea. In 1911, while he was still an actor, Boykow had theorized that an inertial navigational device might be built. By the 1920s he had launched a company to develop aircraft autopilots. In the meantime, he continued research on his automatic navigator. A prototype tested in 1930 fell far short of perfection; indeed, three hours out from the airport in Berlin, the machine indicated that the airplane had reached Australia.[17]

In 1934, not long before his death, Boykow was contacted by a bright young engineer named Wernher von Braun, who was working on a German military program related to rockets. These early rockets had a distressing tendency to tumble aimlessly after launch, crashing to earth. Von Braun wanted to know: would Boykow's navigation system keep the German rockets flying straight up? Boykow's device was installed in a German rocket, called the A-3. However, the system didn't work; all the A-3 tests ended in failure. One of Boykow's colleagues argued, quite plausibly, that the basic idea was sound but beyond the limits of the technology available at that time.

The Germans moved on to alternative steering methods for their rocket. The most effective involved a stream of guidance signals sent by radio. Yet German military leaders disliked the idea, for once the enemy figured out the relevant radio frequencies, they could broadcast a jamming signal and send the rockets flying out of control. The rocket designers eventually incorporated a gyro-based system devised by engineers at the electronics firm Siemens. While entirely self-contained and thus unjammable, the system was not a true inertial navigation system. It was merely good enough to ensure that a missile would fly straight up during launch, then arc over and plunge to earth a couple of hundred miles away. A preset guidance fin on the rocket would aim it in the general direction of its rather large target—more often than not, the city of London.

Over the course of World War II, the Germans launched about three thousand such rockets, named V-2. They scarcely made a difference in the outcome of the war, but they killed about five thousand people. Even more than the V-1, the V-2 was a terror weapon. The V-1 cruise missiles were slow and loud. Air defense crews could hear and see them; many were shot down over the North Sea. But the V-2's one-ton warhead fell toward its target from fifty-six miles up, traveling at supersonic speed. There was no warning and no defense.[18]

Brutal as they were, the V-2 attacks did little to prevent Germany's ultimate defeat. Yet the Allies could see the weapon's immense potential. Had the Germans been able to field them sooner and in greater numbers, they might have altered the outcome of the war, especially if they had been able to improve the accuracy of such missiles and the power of their warheads.

Both the United States and the Soviet Union were eager to exploit German achievements in rocketry, even as relations between the two countries degenerated. America had the atomic bomb; the Russians would have their own in a few years, aided by a successful campaign of espionage against their erstwhile American ally. A

V-2-type missile with an A-bomb as its payload would be a weapon of monstrous power.

However, a V-2 could merely loft a one-ton warhead. Little Boy, the atomic bomb dropped on Hiroshima, weighed five times as much and could never have been delivered by such a missile. Just as important, while the atom bombs of the time could level a city, they would be far less effective against military targets made of reinforced concrete and buried underground. Although the Hiroshima bomb generated the force of eighteen thousand tons of high explosive, that was not enough to reliably destroy a hardened military installation, unless the bomb could be placed within a few dozen yards of the target.[19] In a future US-Russia war, the missiles delivering these bombs would be fired from thousands of miles away.

That was the challenge. Pick a spot in Russia, and draw a circle around it with a hundred-yard radius—about the length of a football field. Now design a missile that could lift off from US soil, travel halfway around the planet, and land within the Russian circle. Like the Germans during World War II, the American missile builders were wary of using radio signals to make in-flight course corrections. Not only were such signals subject to jamming, but it would also be extremely difficult to maintain radio contact with a missile over such great distances. The engineers needed an inertial navigation system that could find its own way, with no further human assistance once the missile was launched. It was perhaps the most daunting navigational challenge since the centuries-long quest to accurately calculate longitude. Yet American scientists, as well as Russians working in the opposite direction, would tackle the problem and achieve extraordinary results in a very short time.

They had plenty of German help. The United States and Russia captured dozens of the Third Reich's rocket scientists, and many proved willing to work for their onetime foes. America's German scientists mainly worked at the US Army's Redstone Arsenal in Huntsville, Alabama. There they dramatically improved the accuracy of early American guided missiles.

The most important work on self-contained guidance was performed at the Instrumentation Laboratory at the Massachusetts Institute of Technology and at the Autonetics Division of North American Aviation in Los Angeles. The work at these two laboratories would lead to extraordinary achievements, including the successful voyage of the *Nautilus* and the navigation of spaceships to the moon.

First, however, the pioneers of inertial navigation had to overcome an extraordinary obstacle—the all too plausible belief that such a device could never be built. And not because of a lack of suitable technology. One of the world's leading theoretical physicists insisted that the very laws of nature made inertial navigation impossible.

George Gamow, a Ukrainian scientist who had fled life under Joseph Stalin and emigrated to the United States, would play a major role in developing the Big Bang theory of the origin of the universe and a minor role in the history of navigation. Gamow asserted that Albert Einstein's theory of general relativity ensured that an inertial navigator would never work.

One of the curious consequences of Einstein's theory was the equivalence of acceleration and gravity. The force of acceleration that pushes you back against your car seat when you step on the gas is indistinguishable from the force of gravity that presses your body downward, toward the center of the earth. Being inside an inertial navigation device would be like being strapped into a seat inside a box with no windows. You feel a force pushing you back in your seat. The box may be moving forward. On the other hand, you may be lying on your back, and the force you feel is gravity pulling you down. Without some external information to help you identify the direction of gravity, known to scientists as the "local vertical," you could not know whether you were feeling acceleration or gravity. A set of accelerometers would be similarly confused and could therefore never be relied upon as a navigational guide. Physicists soon realized the implication of this idea. In 1942 an introductory physics textbook declared that Einstein's theory "constitutes a practical difficulty in the blind navigation of airplanes, since it makes impossible the construction of a device to indicate the true vertical."[20]

Gamow went further, writing a paper titled "Vertical, Vertical, Who's Got the Vertical?" mocking the idea of inertial navigation. His contempt for the idea was especially menacing because Gamow sat on a US Air Force advisory panel tasked with studying problems of aircraft guidance. The opposition of such a brilliant man might have strangled inertial navigation at birth.

That this did not happen was partly due to a gregarious, relentless polymath from Missouri named Charles Stark Draper. Born in 1901, Draper entered the University of Missouri at age fifteen and later transferred to Stanford, where he earned a degree in psychology in 1922. After an impromptu cross-country road trip, Draper visited the campus of MIT and decided he liked it so much he remained for sixty-five years, until his death in 1987. At MIT Draper earned a bachelor's in electromechanical engineering and a doctorate in physics. Enthralled with aviation, he tried to enter the US Army Air Corps, but was rejected because he kept getting airsick. Still, Draper earned a private pilot's license and devoted himself to problems of navigation.

One of the payoffs from his research was the Mark 14 gunsight, which Draper designed with help from the Sperry Gyroscope Company and his MIT students. A gun equipped with the Mark 14 did not aim directly at an enemy plane. Instead, a gyroscopic aiming system automatically "led" the target, putting shells where the plane would be a split second after the shot. In 1942, during its first major engagement, Mark 14–equipped guns aboard the battleship USS *South Dakota* brought down thirty-two Japanese planes. More than eighty-five thousand of the gunsights were built by war's end.

By then Draper was hard at work on inertial navigation systems. He was convinced that the relativity problem was beatable, largely thanks to Max Schuler, a cousin of gyrocompass inventor Anschütz. Years before Gamow had raised his red flag, Schuler had devised a shrewd solution.

Schuler conceived of a long pendulum, hanging from a support on the earth's surface and extending to the center of the earth. The support for the pendulum would be attached to a moving vehicle, which could move in any direction on the earth's surface. As it did, the pendulum's far end would remain at the earth's center, and its upper support would remain level relative to the earth's surface. An inertial navigation system attached to such a pendulum would always be correctly aligned with the local vertical. Thus, it could separate the effects of gravity from those of acceleration, allowing one to accurately calculate the vehicle's location.

Although such a pendulum could never be constructed, the same principle could be incorporated in an inertial navigation unit, by having its gyroscopic platform constantly adjust itself so that it remained level at all times relative to the earth's surface. The platform would have the correct vertical alignment at the beginning of any voyage. As the inertial navigation device moved, the curvature of the earth would change the orientation of its gyroscopic platform, but the accelerometers could measure the change and command a set of motors to keep the platform correctly aligned with the center of the earth at all times.

The technique became known as "Schuler tuning," and subsequent experiments confirmed its effectiveness. It was not a perfect solution, because the planet's gravitational field is far from uniform, but improved measurements of the earth's gravity field could compensate for this variation. Schuler tuning overcame Gamow's conviction that inertial navigation was theoretically impossible—it was merely very difficult.

Armed with this knowledge, Draper arranged a top-secret conference with the leading researchers in inertial navigation and invited Gamow to attend. Draper believed that evidence of progress in addressing the problem would outweigh any theoretical quibbles. In the end, Gamow never played his Einstein card; he did not even show up for the meeting. The military took this as a tacit admission of error and continued to invest in inertial navigation research.

The payoff would be a while in coming. The accuracy of a human navigator's dead reckoning depends on the precision of his measurements of time, speed, and distance traveled. Similarly, the accuracy of an inertial navigation system is limited by the precision of its accelerometers and gyroscopes. The researchers at Autonetics and MIT wanted a system that would arrive within a mile of its intended target after flying for ten hours—the flight time of a bomber heading from the United States to Russia. In 1947 Draper calculated that this would require gyros one hundred times more precise than anything then available.

The problem gradually succumbed to dogged effort. At first Draper and the Autonetics team compromised, building semiautonomous navigators that corrected themselves using old-school stellar navigation. These machines included optics that could identify specific stars and use the data to get a better fix on their location.[21] Meanwhile, engineers addressed the imperfections of the gyroscope, developing versions that floated on cushions of pressurized gas or in pools of thick, electrically warmed liquids. The devices were so sensitive that the slightest contamination could ruin them, so they came to be assembled in dust-free "clean rooms," early versions of the kind used in today's microchip factories. Slowly and steadily, the gyros began to meet the standards Draper had set. It became possible to build pure inertial navigation systems that did not need any external assistance.

In February 1953 a B-29 bomber lifted off from Hanscom Air Force Base in Bedford, Massachusetts, for a nonstop flight to Los Angeles. The navigator wasn't human. For twelve hours and 2,250 miles, the aircraft was steered by Space Inertial Reference Equipment, or SPIRE, a twenty-seven-hundred-pound product of Draper's MIT lab. The first full-up test of a pure inertial navigator was a complete success. Apart from a single preplanned course correction, SPIRE found its way without assistance—no radio inputs, no star sightings. Draper, the master engineer, had refuted the great theoretician Gamow, once and for all.

The SPIRE test was conducted in secrecy, with good reason. Its military implications were immense. Here was a self-contained system that would be ideal for long-distance navigation of ships and airplanes. As the voyage of the *Nautilus* would demonstrate nearly six years later, an inertial system could guide a submarine for thousands of miles under water, steering it to within a few miles of its intended destination. The same technology would work for surface warships and long-range bombing planes as well.

Above all, inertial navigation enabled the construction of missiles capable of hitting a target thousands of miles away. Germany's V-2 rockets were accurate only to within 4 miles, after having traveled a mere 200 miles.[22] A missile fired from the heart of the United States to Russia would travel more than 5,000 miles; at the V-2's standard of accuracy, the missile might easily be more than 100 miles off target. Even with nuclear warheads, that's lousy shooting. An inertial system could reduce that error dramatically, to within a mile or two of the target even after flying halfway around the world. Such accuracies were not possible right away, but by the 1970s the United States and the Soviet Union were building missile guidance systems that, on paper, were believed to be accurate to within 900 feet.[23]

It was a remarkable accomplishment, and, to some, a terrifying one. Far from making the world a safer place, ultra-accurate ICBMs (intercontinental ballistic missiles) might increase the risk of nuclear war, by encouraging US or Soviet politicians to suppose that they could destroy the other's missiles in a preemptive strike. Older missiles were not accurate enough to take out hardened missile silos, so there would be no point in trying. Newer models, however, like the American Minuteman III and MX missiles, or the Soviet SS-18s, could have done the job, leading some planners to suggest that a US or Soviet first strike might just work.[24]

Whether through shrewd leadership or good luck, the Soviet Union collapsed, the Cold War ended, and the worst never happened. However, the technology to eliminate remote enemies had been launched and would provide a template for later applications, most recently in the controversial use of drones to attack terrorist strongholds in the Middle East.

At the same time, inertial navigation was applied to a host of non-military tasks: for instance, Draper's laboratory at MIT built inertial navigation systems for the Apollo program. At around the same time, inertial systems were coming onto the civilian market. They had become much smaller over the previous decade and a good deal cheaper—a mere one hundred thousand dollars or so. This was good news for commercial airlines. For years they had been looking for ways to save money by cutting down on the size of flight crews. In the early days of jet travel, planes routinely carried a crew of four—pilot and copilot, a flight engineer who managed the aircraft's engines and other mechanical systems, and a navigator to point them in the right direction. International air regulations required the presence of a navigator on aircraft making transoceanic flights. As aircraft became more reliable and more automated, the job of flight engineer gradually became superfluous. These days the few still on the job fly for cargo carriers that operate older airplanes.

Airlines began replacing navigators even earlier. In the early 1960s, Trans World Airlines began installing Doppler radar on its planes. This new type of radar could measure the plane's location based on the radio echoes from the ground below, and it was simple enough that the pilot and copilot could use it themselves. Its invention eliminated hundreds of navigator jobs.[25] Inertial navigation would take care of the rest.

The Boeing 747 became the first commercial jet to install an inertial rig as standard equipment—three of them, in fact, with two as backups. The system, called Carousel and manufactured by the Delco Electronics division of General Motors, let the pilot and copilot punch in the desired course as a series of waypoints, like bread crumbs in space. The Carousel then steered the plane to its destination, hitting one waypoint after another. The pilot and copilot did little more than watch.[26]

The early commercial systems, like those used by the military, relied on traditional spinning gyros. Eventually, this technology would be supplanted by a new kind of gyro that uses two beams of laser light in place of a rotating disk. The frequencies of the two beams are slightly altered by changes in the vehicle's motion, producing an instant digital measurement of the alteration.[27] At first less reliable than traditional gyroscopes, these ring laser gyros have since become the standard for inertial navigation systems.

With inertial navigation, humans had invented the first fully automated wayfinding technology. Although it was good enough to steer submarines to the North Pole, jumbo jets across the Pacific, or nuclear-tipped missiles to Moscow, it still was not perfect. As with dead reckoning performed by human navigators, tiny errors inevitably creep into the calculations of an inertial navigation system, making it less and less reliable over time. Inertial navigation was good enough for most purposes, but for even greater precision scientists once again looked to the skies—and an unexpected gift from a triumphant enemy.

In Transit

ON A CLEAR AUTUMN NIGHT IN 1957, YOU COULD SEE IT IF YOU KNEW where and when to look. Up, of course, and far above your head. With help from newspaper and magazine accounts, you could choose the right time of night for looking and guess the right piece of sky to stare at. If you got it right—and many people did, all over the world—you would see a glint, racing across the blackness.

You would get a good deal more if you owned a radio, the kind used by serious amateurs. Hams all over the world barely bothered to look up; many never left their basements, attics, or radio shacks. Instead, they tuned with delicate fingers until they had dialed up the correct frequency—20 megahertz.

It was the frequency used by WWV, a government-run station in Greenbelt, Maryland, that broadcast nothing but a precise time signal, twenty-four hours a day, seven days a week, every day since 1920. Scientists, businesses, armies, and navies set their clocks by WWV—that's what it was there for. In late 1957 there was something more to listen for at 20 megahertz; at certain hours WWV would fall silent for a few minutes, to let Sputnik have its say.

The world's first artificial satellite, launched by the Soviet Union on October 4, chirped a rhythmic electronic beep, once every second, for twenty-one days until its batteries failed. Sputnik itself was too small to be seen by the naked eye; the gleam spotted by earthly viewers was sunlight reflected off the one-hundred-foot rocket that had boosted the satellite into orbit.

Republican president Dwight Eisenhower responded, at first, with a shrug. The United States was hard at work on its own satellite program, set to make its first launch in 1958. So the Russians got there first. Yet angry Democrats in Congress did not see it that way, nor did fretful newspaper columnists. Within a few days the mood of the nation had shifted to something not too far from panic. Dismayed citizens gaped at the gleaming booster and imagined a near future in which the Soviets might orbit something far more deadly. America's mortal enemy had seized the high ground.[1]

Not even the Russians fully grasped what they had accomplished. They certainly had no idea that they had created a rudimentary method for humans to accurately determine their location on earth. Sputnik wasn't only the first satellite; it was the first navigational satellite. It was not what its Russian creators had intended, but a band of enterprising Americans were quick to walk through the doorway that the Soviets had unwittingly opened.

The day of Sputnik's launch was a Friday, giving the world a weekend to mull over the Soviet scientific milestone. When they returned to work the following Monday, William Guier and George Weiffenbach could scarcely talk about anything else.[2]

Guier, who held a doctorate in theoretical physics from Northwestern University, and Weiffenbach, finishing up a physics doctorate from the Catholic University of America, both worked at the Applied Physics Laboratory at Johns Hopkins University in Laurel, Maryland. Founded in the aftermath of the Japanese raid on Pearl Harbor, the APL's scientists and engineers specialized in practical research to benefit the nation's security. The fact that a nuclear-armed Russia could launch objects into orbit over the American heartland should have captured the APL's full attention. Yet none of the lab's scientists had come in over the weekend to consider such a problem. For the past three days, Sputnik had been beeping away, but the APL had not been listening. "The more we discussed the issue, the more keen we became on listening in," Guier and Weiffenbach recalled more than forty years later.

Weiffenbach had an ideal radio receiver for the purpose. By tuning to 20 megahertz, the two scientists easily pulled in the satellite as it passed overhead and picked up the time signal from WWV, based just twelve miles away in Greenbelt. Twenty megahertz is far too high-pitched for our ears to detect, so the scientists used a device called a beat-frequency oscillator to downshift the signal to within human listening range. On a few evenings WWV briefly interrupted its time-signal broadcasts to let amateur radio operators hear the pure Sputnik signal. For the most part, the satellite and the earth-based transmitter operated simultaneously, making it easy for scientists to make tape recordings of the two signals as they overlapped. However, they did not overlap exactly. The Russian signal sounded a little off-key compared to the precise, unvarying sound of WWV. And whereas WWV was anchored to the Maryland soil, Sputnik swooped overhead at eighteen thousand miles per hour.

This discrepancy was caused by the Doppler effect, the way that motion stretches frequencies. Just as the sound of a locomotive's air horn sounds consistent to the engineer traveling on the train but warps to a listener standing at a fixed point next to the track, radio waves from a moving object also shift in frequency.

The APL had proven the usefulness of the Doppler effect during World War II, through its development of the proximity fuse, described in Chapter 2. Each proximity-fused shell carried a small transmitter that bounced radio waves off its target. The shell also carried a simple receiver to pick up the reflected radio waves. Thanks to the Doppler effect, the frequency of those reflected waves would shift upward as the shell got closer and closer. If the shell missed and flew past the target, the frequency would begin to shift downward. The

proximity fuse instantly detected this shift and exploded the shell—
bringing down the plane without a direct hit.

So when Guier and Weiffenbach noticed that the frequency of
the Sputnik signal shifted anywhere from 500 to 1,500 hertz as the
satellite passed, they realized they could calculate the orbital path
of the satellite with great accuracy, by measuring the Doppler shift.
Before their work was done, however, the radio on Sputnik went dead.
In November the Russians launched a second satellite, Sputnik II.
Better yet, this one broadcast on two different radio frequencies at
once. This helped the scientists calculate the amount of distortion
that occurs when radio waves pass through the ionosphere, an atmospheric layer full of electrically charged particles. By factoring out this
ionospheric distortion, APL scientists could get a more accurate fix on
the satellite's location. Eventually, the scientists at the APL were able
to predict when they would next pick up Sputnik II's signal—proof
that they had accurately calculated the satellite's orbit relying solely
on its radio transmissions. It was a worthy achievement and proof that
the Russians were not the only ones deriving scientific benefits from
Sputnik's launch.

However, there was more to come. On March 17, 1958, Guier and
Weiffenbach were called to the office of Frank McClure, director of
the APL research center, to answer a question. If it was possible to
calculate the position of Sputnik from its radio signal, McClure asked,
couldn't you use the same signal to identify the location of a radio
receiver on earth? In hindsight it seems an obvious idea, but it had
not occurred to Guier or Weiffenbach. Still, they threw themselves
at the problem.[3] "We did not fully realize the potential of what we
were doing," they said. McClure, however, did. The Canadian-born
chemist was an expert on solid rocket fuels for guided missiles. His
expertise made him welcome at a new branch of the US Navy, the
Special Projects Office (SPO), which sought ways to put guided missiles aboard ships at sea.

The German V-2 rockets, which the United States had captured
during the last days of World War II, were state of the art in 1940s

guided-missile technology. As mentioned in Chapter 3, however, they were not powerful enough to carry the massive atom bombs of the era. The plutonium bomb that leveled Nagasaki weighed ten thousand pounds. For years after the war, the US military did not see much future in long-range missiles, focusing instead on developing massive new bombing planes capable of lugging nukes to Moscow.

This mind-set began to change in the early 1950s with the development of relatively lightweight hydrogen bombs and more powerful rockets to carry them. The Eisenhower administration committed to building a fleet of nuclear-tipped missiles, capable of destroying cities thousands of miles away. With billions in federal funding on the line, the US Army, Navy, and Air Force all wanted a share of the spoils. By 1955 President Eisenhower had decided that the nation should have four separate guided-missile programs. The air force won approval for three of them, and the army was favored to win the fourth.

All these missiles were to use liquid fuel, like the old V-2s. Yet the navy wanted no part in liquid-fueled rockets. You could certainly launch a liquid-fueled rocket at sea; the navy had fired a V-2 from the aircraft carrier *Midway* back in 1947. Still, liquid fuels were a shipboard nightmare. They ate through metals, cast off toxic fumes, and could spawn massive fires or explosions. A 1948 test that simulated a shipboard V-2 fuel explosion terrorized the navy brass.[4] For years thereafter, they focused on developing cruise-missile technology modeled after the German V-1s. Basically a pilotless jet aircraft, a cruise missile is far slower and has less range, but is less likely to get its users killed.

In 1955 newly appointed chief of naval operations Arleigh Burke took a different view. Burke created the Special Projects Office to work with the army on its liquid-fueled Jupiter. At the same time, the navy began developing a solid-fueled version for safer shipboard deployment. In 1956 SPO officials concluded the navy could do better. They won permission to develop an entirely new missile called Polaris. It would be solid fueled, smaller and lighter than the Jupiter, yet capable of being launched from a submerged submarine. The Polaris could lift a hydrogen device thirty times more powerful than the Hiroshima bomb and drop it up to twelve hundred miles away.[5]

McClure knew all about the new missile program; he had led an SPO panel that helped develop a safe, smooth-burning solid fuel for the rocket. McClure was also aware of a crucial problem in launching a long-range missile at sea—how could it find its way to the target? The V-2s that had battered London were dumb terror weapons, designed to carry a one-ton high-explosive bomb no more than two hundred miles. As long as its crude gyroscopic guidance system sent a V-2 in the right general direction—across the English Channel—and dropped it somewhere on British soil, the German engineers could call it a win.

Sending a missile on a thousand-mile journey was another matter. As the range increased, so did the need for accuracy in targeting—a one-degree targeting error would cause the missile to be a mile off course after traveling only sixty miles. The farther the missile traveled, the farther off course it would go. By the end of a thousand-mile journey, the warhead would land more than sixty miles off target. Given the hellish destructive force of an H-bomb, naval planners did not insist on pinpoint precision. Landing within two miles of the target was considered acceptable. Beyond that, they could not be sure they would completely destroy the selected target, even with a nuclear weapon.

The navy already had a general idea of how to guide the weapon, because they had faced a similar problem navigating their new fleet of nuclear submarines. Conventional subs spent more time on the surface than below it, so they could rely on the same well-understood navigational techniques used by other ships. However, the point of a nuclear sub is that it scarcely ever needs to surface. Nuclear reactors need no air to produce electricity to drive the boat's power plant and electronics, and there is plenty of energy left over to make fresh oxygen and drinking water from the sea itself. With inertial navigation to steer the boat, a nuclear sub could remain submerged for months, until the food and the crew's patience ran out.

Inertial navigation, however, while acceptable for a submarine, would not work accurately enough for a missile. Polaris missiles would have no hope of achieving their desired accuracy, unless each missile's inertial navigation unit could be programmed with the starting point of its flight. This is a trivial matter with land-based intercontinental ballistic missiles; the location of each silo is measured down to the square inch, and it never moves. On the Polaris boats, the missile guidance systems would be fed location data from the sub's inertial navigational system. And after weeks or months at sea, the sub's location fix would not be nearly accurate enough to support a missile launch, thanks to all those accumulated imperfections of the inertial navigational unit.

Of course, the sub could surface and try to get a location fix from sun, moon, or stars, but a surfaced sub is an easy target for enemy attack. The navy developed a periscope that could poke out of the waves and let skippers take star sightings while the bulk of the submarine stayed hidden. Still, the navy found that periscope sightings offered "marginal accuracy," while the periscopes were costly to maintain. And as with all such navigational methods, it was rendered useless by cloud cover. There were also land-based radio navigation aids like LORAN-C, which could be detected by merely poking a radio antenna above the waves. However, the LORAN network was not global; indeed, it was often unavailable in the northern latitudes where Polaris boats would patrol.[6]

In other words, the navy needed a whole new way for a ship to pinpoint its location. This was the problem that McClure's friends in the navy were sweating over. And he realized that a pair of his researchers might have stumbled across the answer. McClure suggested to Guier and Weiffenbach that if they could figure out the location of the satellite based on the Doppler shift of its radio signal, it should be possible to work the same calculation in reverse. Once the satellite's position was known, someone on the ground or the surface of the sea, equipped with a radio receiver, would be able to figure out his own position to within a couple of hundred feet by tuning in on the satellite as it passed overhead. A submarine equipped with such a system could regularly update its inertial navigational system by sticking an antenna out of the water.

Even a brief antenna exposure could make a submarine vulnerable to radar detection. But the sub could raise the antenna at times of its own choosing, using sonar to ensure there were no enemy warships nearby. After a few minutes of listening to the satellite, the sub's navigators would have a very accurate idea of the boat's location, suitable for aiming Polaris missiles at Moscow. McClure's idea took Guier and Weiffenbach by surprise—it had simply not occurred to them. Yet something similar had occurred decades earlier to someone in no position to do anything about it.

In the years following the American Civil War, humans were decades away from learning to fly, much less from putting objects into orbit. Still, it seemed like a good idea to Edward Everett Hale, a Unitarian clergyman and chaplain to the US Senate. Hale was a popular writer of short stories, and in "The Brick Moon," written in 1869, he made a prescient suggestion.

> If, from the surface of the earth, with a gigantic peashooter, you could shoot a pea upward from Greenwich, aimed northward as well as upward; if you drove it so fast and far that when its power of ascent was exhausted, and it began to fall, it should clear the earth and pass outside the North Pole; if you had given it sufficient power to get it halfway around the earth without touching, that pea would clear the earth forever. It would continue to rotate above the North Pole, above the Feejee Island place, above the South Pole and Greenwich, forever, with the impulse with which it had first cleared our atmosphere and attraction. If only we could see that pea as it revolved in that convenient orbit, then we could measure the longitude from that, as soon as we knew how high the orbit was, as well as if it were the ring of Saturn. "But a pea is so small!" "Yes," said Q., "but we must make a large pea."[7]

Hale had no notion of radio; he simply proposed that people on earth could get their bearings by observing the position of what he called the Brick Moon. Still, Hale was probably the first person to think of using artificial satellites as navigational aids.

Nearly ninety years later McClure, Guier, and Weiffenbach were ready to try it. "This concept may offer the Laboratory an ideal opportunity," McClure wrote in a memo to APL director Ralph Gibson. "First, it definitely puts the Laboratory into the space game; second, it offers a problem that has real military significance; third, it is definitely a Navy problem and therefore fits in well with the Laboratory's long-time Navy relationship; and fourth, it is of direct significance to SP [the Special Projects Office], which is the only activity in the Navy presently with which we have close contact which has funding sufficient to support such a project."[8]

Richard Kershner, an APL mathematician who had been working with the navy on the Polaris missile system, was put in charge of developing the satellite navigational system. Kershner and his team came up with a plan for satellites that would maintain a circular orbit over the North and South Poles, with each complete orbit taking about 108 minutes. A network of ground stations would use the techniques pioneered by Guier and Weiffenbach to calculate the exact orbital path of each satellite and beam that data up to the satellite, which would store it in an electronic memory. Each satellite would constantly broadcast its orbital data over two different frequencies. A ship or submarine with a receiving set would pick up the broadcasts and feed them into a computer. Using the two frequencies would enable the computer to cancel out ionospheric distortion and get a clear reading on the Doppler shift of the satellite's signals. The orbital path data included in the broadcast would tell the computer the satellite's location; the shape of the Doppler shift, measured over several minutes, would reveal the submarine's position relative to the satellite. Kershner's team estimated that the system would be accurate to within one-tenth of a nautical mile—about six hundred feet. And if a Polaris missile's guidance system "knew" its launch point with that degree of accuracy, it could be counted on to deliver its warhead to within a half mile of its intended target.

On April 4, 1958, APL director Gibson pitched the idea to the US Navy's Bureau of Ordnance. The navy was interested, but not interested enough to pay the $1 million needed to start building satellites. In December the money came through, but not from the navy. Instead, the check came from the Defense Department's Advanced Research Projects Agency, or DARPA—the same outfit that a few years later provided the seed money for the Internet.

By now the satellite program had a name—Transit—and a plan to loft three 270-pound satellites into polar orbits beginning in August 1959, less than two years after Sputnik. Although there were delays in the schedule and one satellite was shattered during a late-July laboratory test, the first of the series, Transit 1A, was mounted atop a Thor-Able rocket at Cape Canaveral on September 17. The countdown went to zero, the rocket left the pad, and the transmitters aboard Transit 1A hummed out their signals as the satellite soared. Then everything went dead. Transit 1A had disappeared, victim of a defective third-stage booster that sent the satellite plunging back to earth. It was a disappointment, but hardly a surprise. It was 1959; in those days, a lot of satellite launches went bad.

By April 13 the Transit team was ready to try again with Transit 1B. The satellite made orbit, although at a rather lower altitude than the scientists had hoped for. Tracking stations in Texas, New Mexico, Washington State, Canada, and Great Britain picked up the reassuring radio signals that proved the satellite was functioning properly. The only problem was that when the scientists tried using the Guier-Weiffenbach method to pinpoint the satellite's location, they found that their results were routinely off by two or three miles. "The satellite was all over the sky," said Guier.

The APL team had neglected to allow for the vast variations in our planet's gravity field. To a human traveling between cities or continents, gravity may feel the same everywhere on earth, but it isn't. Quirks in the world's shape and geological structure lead to variations in the strength of the gravity field. It soon became clear that space scientists would need accurate gravity maps of the planet before they would be able to correctly predict the orbits of satellites—or the reentry paths of nuclear warheads, for that matter.[9]

Transit had been planned to be operational by 1962, two years after the first Polaris-armed nuclear submarine put to sea. In reality, it would take considerably longer to solve the gravity problem. Four more Transit satellites went up between June 1960 and November 1961. These were prototypes that were not intended to provide navigational aid to navy ships. Instead, they were mainly used to map the world's gravitational irregularities. Isaac Newton had long ago surmised that, far from being perfectly spherical, the earth had an oblate shape—flatter at the poles and thicker at the equator than a perfect sphere. Transit observations confirmed and refined Newton's theory. Researchers who had studied the orbit of an earlier US satellite, Vanguard I, concluded that the earth was actually pear shaped, with the Northern Hemisphere somewhat longer than the Southern. The effort to predict the track of Transit 1B corroborated this, and factoring in the gravitational effect of a pear-shaped world produced a much more accurate result. By late 1964 Transit researchers had mastered their craft. They could now calculate the location of one of their satellites to within about three hundred feet. That information, beamed up to the satellite and then back to earth, allowed for extremely accurate navigation.

The first fully operational Transit satellite went up in late 1963 and was put to work by the navy the following year. Unfortunately, Transit 5BN-2 died after just a few months in orbit. This would prove to be a recurring problem; indeed, the history of the early Transit satellite launches is a litany of failure. Transit 5BN-3 crashed back to earth during launch. As for the next series of Transit satellites, code-named "Oscar," the first nine failed within months or even weeks of launch, while a tenth was only partially operational. APL engineers blamed the Naval Avionics Facility in Indianapolis for poor construction. The APL took over construction of the next batch of Oscar satellites; later, RCA won the contract to build them. Between them the APL and RCA did something right; their Oscars had an average life span of fourteen years in orbit.

In 1967 the US government declassified Transit and granted permission for commercial use of the technology. The following year companies like Magnavox were selling Transit receivers for use on merchant ships at about fifty thousand dollars each. By 1968 there were four Transit satellites in orbit, enough to ensure that at least one was in range of most ships most of the time. Eventually, the system used a "constellation" of five satellites.

In 1981 Magnavox introduced the MX4102, a Transit receiver cheap enough for use by amateur sailors. Five years later retired US Marine Teddy Seymour took one along on a solo around-the-world cruise, the first by a black man. "Now, there are two things I won't leave port without," Seymour later wrote. "Satellite navigation and M&M's with peanuts."[10]

Transit had been invented by the United States as a military technology, but having made it available to the world, any of the world's armies or navies could take advantage of it. When Argentina and Great Britain went to war over control of the Falkland Islands, the navies of both countries used Transit as a navigational aid.[11]

Although Transit was an extraordinarily effective technology, it suffered from a couple of serious limitations. First, Transit satellites were not always within range. A ship at sea might have to wait for several hours before one of the Transit birds passed overhead. This was not a fatal problem for a submarine that needed to adjust its inertial navigation unit every week, but it meant that the satellites were no good for providing constant navigational updates. In addition, Transit was a two-dimensional system. From its Doppler shift, a receiving unit could calculate latitude and longitude, but not altitude. A user would know how far north, south, east, or west he was, but not how high above the land or sea.

Although it was not a concern for sailors, altitude is extremely important for calculating the trajectory of a guided missile. Transit improved accuracy by letting sub skippers program their missiles with their exact location at launch. Yet a missile could be made even more accurate if it could constantly recalculate its location in flight. To do so it would need a stream of signals that would provide three-dimensional navigational data. Such a missile would be capable of landing within a few feet of its target even after traveling halfway around the world.

Aircraft pilots would also welcome 3-D satellite guidance. Transit's two-dimensional technology held little appeal for the aviation industry. A satellite system that measured latitude, longitude, and altitude would make an excellent supplement for land-based radio navigation systems like LORAN.

The next leap forward in satellite navigation would be prompted by one of the Cold War's most bizarre weapons programs, a plan to put rockets on rails. Worried that the Soviets were developing missiles powerful and accurate enough to hit America's land-based missile silos, the US Defense Department around 1960 considered building a new generation of solid-fueled Minuteman ICBMs, small enough to fit inside specially modified railroad cars. These railcars would be dispersed throughout the US rail network during periods of maximum peril, when the president might judge the risk of a Soviet surprise attack unusually high. With millions of railcars scattered over thousands of miles of track, the Russians would never be able to get them all. They would be a terrestrial hide-in-plain-sight version of the navy's submarine-launched missiles, inaccessible and hence indestructible.

However, these mobile Minuteman missiles would have to be aimed. Unlike their fixed counterparts, they might be fired from any of a million locations, and that location data would have to be programmed into each missile before launch. It might take an hour or more to get a Transit fix—too slow. The railcar might be in the mountains or on a high plateau, almost certainly not at sea level, and so elevation data—that third dimension that Transit could never provide—would be essential.[12]

Such a plan would require a major improvement in navigation, a technology that could provide a highly accurate three-dimensional location fix within seconds. This navigational system was a far better idea than the missile plan that spawned it. So although the Kennedy administration abandoned the idea of nuclear bombs in boxcars, work began on a space-based navigational technology that would make Transit obsolete.

Two exceptional engineers tackled the project, one in the US Navy and another in the Air Force. Four decades later the two men would clash over who deserves the lion's share of credit for today's space-based navigational systems. But for hundreds of millions throughout the world, the important thing is their newfound ability to pinpoint their location anytime, anywhere—especially when their lives depend upon it.

Found in Space

THE SOUND OF THE AIRPLANE SAVED HER LIFE. IT WAS JANUARY 1997, and Karen Nelson, a fifty-one-year-old cook at a South Dakota nursing home, had made the mistake of trying to drive home through a massive blizzard. Her car swamped by the rising snowdrifts, Nelson was trapped inside the vehicle as the outside temperature fell to twenty below zero. She had blankets, a sleeping bag, and a cell phone. Her call for help got through to local police. Help was on the way—or it would have been, if anyone had known where to look. Nelson had only a vague idea of her location; local landmarks were disguised beneath the snow. For all its virtues, Nelson's phone lacked any means of pinpointing her exact location. No one's phone could do that, back in 1997.

Today, nearly any phone can. For hundreds of millions around the world, our phones have become personal navigators that can guide us accurately for thousands of miles or pinpoint our exact location at a moment of life-or-death crisis. Frightened, desperate people like Nelson helped make it happen.

Nelson spent forty hours in her car, powering her phone with help from the vehicle's fading battery. Above her, a search airplane crisscrossed the snowy landscape; Nelson spoke with the pilot. He suggested that she listen for the sound of his engine and flick her headlights on when she heard him pass. In time, she heard the drone, hit the lights, and watched them flicker on and then die for good. The battery was dead. However, Nelson's timing was excellent. She had been spotted, and snowmobilers soon carried her to safety.[1]

Nelson was fortunate. For other lost souls with early cell phones, help arrived too late. In January 2000 Terry Allen Pedigo collapsed during a soccer game in Houston. Pedigo's son grabbed his father's cell phone and dialed 911. There was a fire station a mile and a half away; they could have been on scene in five minutes. Yet the 911 dispatcher had no way of knowing this—Pedigo's son did not know the street address of the park where his father lay. In the end, help arrived in twelve minutes, a fairly rapid response but not quick enough to save Pedigo's life.[2]

The poignant tale of Karla Gutierrez made national headlines. In February 2001 Gutierrez accidentally drove her car into a water-filled canal alongside the Florida Turnpike. Gutierrez grabbed her cell phone, dialed 911, and begged for help. However, she had no idea where she was. For three minutes, as the car sank into the canal, the dispatcher questioned Gutierrez, probing for clues to her location. And then silence. The cops who scoured the area came across skid marks where Gutierrez's car had gone out of control. They found the vehicle under water, fifty minutes after the call had come in.[3]

In 1997 about 55 million Americans had a cell phone, and that number was growing by 25 percent a year. Eventually, most Americans would rely on them—would bet their lives on them. In the late '90s, however, they would have lost that bet, because there was no sure way of locating a caller.

At that time most landline phone systems used digital switching systems that could instantly identify the source of a phone call. Dial 911, and the call center immediately received the caller's phone number. It was fed to an Automatic Location Identification, or ALI, database, which provided the physical address of the phone. Even if the caller was too sick or injured to give his location, the rescuers could find him.

Not so with cell phones. By definition, they and their users are constantly on the move, so a cell phone number does not correspond to a fixed physical location. Yet every cell phone is a little homing beacon that gives the phone company a rough idea of the phone's location. That is why your cell phone works wherever you take it. When it is switched on, the phone sends signals that are received by the nearest cell tower. The phone company's computer network is informed that your phone is currently within range of a particular cell tower on State Street in Chicago or near the Empire State Building in New York. When there is a call for you, it is routed to the correct tower in the correct neighborhood.

However, the network determines only your approximate physical location. That's good enough for an incoming call from your kid, but useless for bringing you instant aid in an emergency. If your call for help is routed through a cell tower a half mile from where you are dying, finding you in time will be a matter of luck.

Yet everyone knew the problem could be solved. We had seen proof of it in 1991, in the deserts of Iraq. Desert Storm was the military debut of the global positioning satellite system, a new network of orbiting radio stations whose signals allowed soldiers to navigate across barren terrain with extraordinary precision. Civilians also used GPS; anybody could buy a receiver at a sporting goods store for a few hundred bucks.

If you could shrink the underlying electronics so that they fitted on a single silicon chip, then GPS would become so cheap that you could add the technology to any phone for a few dollars. You would then have a phone that would always know its own location, that could not only call for help but draw a road map for your rescuers. For future Karen Nelsons, help would arrive in minutes instead of hours.

By the mid-1990s thousands of hikers and backpackers relied on handheld GPS devices to navigate the wilderness. In-car GPS navigation systems first appeared in 1995 and gradually became a commonplace in luxury cars. However, it was the decision to build GPS into hundreds of millions of commonplace phones that transformed the technology from a costly curiosity to an everyday necessity. For the first time in human history, anybody could own a cheap device that would tell her exactly where she was and exactly how to get where she wanted to go. This constant, instant awareness of our exact location has indelibly altered the way we live and work and travel.

As with inertial navigation and the Transit satellites, GPS was born of the US military's Cold War quest to deliver devastating firepower to exactly the right spot, anywhere on earth. Transit had come first, but its limitations were all too obvious. The five-satellite system did not provide constant coverage of all parts of the planet. Furthermore, Transit provided only latitude and longitude information. That was not good enough to aim the mobile Minuteman rockets that the air force wanted to deploy on railcars scattered throughout the United States. To aim them properly, their launch crews would need a three-dimensional location fix—latitude, longitude, and altitude above sea level.

To accurately calculate all three would require more than one radio signal. A brilliant MIT-trained engineer named Ivan Getting, who had worked on the antiaircraft gun radars that had shot down so many German V-1 cruise missiles, figured out a solution. In 1960 Getting was vice president for engineering and research at defense contractor Raytheon Company. He and a colleague, Shep Arkin, designed a method that used four different radios located at fixed locations, transmitting a precise time signal. Radio waves travel at roughly the speed of light; even so, it takes time for a signal to cover thousands of miles. A receiver could read the radio time signal and then calculate the difference between that time and the actual time the signal had been received. Two radio signals would allow calculation of latitude and longitude; a third would make it possible to measure altitude. There would be tiny, unavoidable errors in the time signals from each radio. Getting and Arkin found that using a fourth radio signal would let them compute the amount of these errors and filter them out, making the resulting location fix far more accurate.

The Getting method had other big payoffs. Transit's single-satellite method was available only when the satellite was passing overhead, and it was time-intensive. The new method would use a network of radios, which would be accessible at all times. Also, the time needed to calculate a location fix would be dramatically reduced; even after a missile was launched, it could use the incoming radio signals to home in more precisely on the target.[4]

Getting called the system MOSAIC, for Mobile System for Accurate ICBM Control. It impressed the air force leadership, but their decision to abandon the mobile Minuteman scheme put the idea on ice for a while. Not long thereafter, Getting left Raytheon to launch the Aerospace Corporation of El Segundo, California, a federally funded nonprofit company that carried out leading-edge space research for the air force. The Aerospace Corporation would tackle a host of demanding projects, from the design of intercontinental ballistic missiles to the development of key technologies for putting men on the moon.

Although busy with his new venture, Getting was still keen on his MOSAIC concept—if anything, even keener. The original idea had featured land-based radio transmitters. Yet the same approach could be even more useful if the radios were placed instead on satellites orbiting the planet. Getting imagined a constellation of satellites whirling above the earth, so many of them that the people below would always be within range of at least four of them. As with Transit, each satellite would receive regular radio updates from ground stations that had calculated the satellite's exact location. These satellites would beam this location data back to earth, along with the exact time of day. A properly designed receiving set could use this incoming data to calculate its own location on the ground. Because the satellites would never be out of range, it should be possible to get a position fix in a few minutes, anytime and nearly anywhere on earth. With four satellites in range, the fix would always include altitude data. Such a system could be used to navigate aircraft as well as surface ships and ground troops.

It was little more than a clever idea at this point, but one that the air force was willing to explore. In 1962 the service contracted

with Aerospace Corporation to launch Project 621B, an in-depth research program to design a new space-based navigational system. It
was the first step in the creation of today's GPS, the beginning of the
beginning.

The air force was late to the party. By the time Project 621B was
launched, the US Navy had already put the first of its Transit satellites in orbit; the system became operational two years later, in 1964.
Navy scientists, however, were well aware of Transit's limitations—
none more so than Roger Easton, an engineer at the Naval Research
Laboratory in Washington.

Easton, like Getting, was an exceptional scientist. By this point he
had already helped develop Minitrack, a system for tracking the location in space of satellites launched by the navy's Project Vanguard.[5]
He had also led the design of the Naval Space Surveillance System,
a system capable of tracking the location of nearly every man-made
object in orbit.[6] Still in use today, the space surveillance system is
presently operated by the US Air Force.

In 1964 Easton proposed a major upgrade to Transit. Instead of
four or five satellites, he proposed a couple dozen, lofted into orbits about eight thousand miles high. Each orbit would take about
eight hours, and the network of satellites would be sent into several
different orbital planes, so that several satellites would be overhead
at any given moment, practically anywhere on earth. Easton's most
innovative idea was to include atomic clocks on each of the satellites.
These timepieces had never been carried into space before. Yet for
the kind of navigational precision the navy sought, atomic clocks
were essential.

Despite its evocative name, there is nothing radioactive about an
atomic clock. But these clocks have a lot of atomic action going on,
down at the level of quantum physics. That is where scientists, in the
1940s, went looking for something they had never possessed before—
an objective and absolute measure of time. Up to then, humans had
measured time by the rotation of the earth. A day was the time it took
for the earth to make a full rotation around its axis; an hour 1/24th

of a day, a second 1/3,600th of an hour. In effect, the spinning earth itself was a giant clock.

Yet by the late nineteenth century, scientists like the Canadian-born American polymath Simon Newcomb had begun to realize that the planetary timepiece was slowly winding down, like an old-school pocket watch. The gravitational force of other celestial bodies, such as the moon, constantly pulls on our world, creating our oceans' tides in the process. The same force gently taps the brake pedal on our planet. As a result, the days get longer, and not only in the springtime. It takes the earth today about two milliseconds longer to spin on its axis than it did two hundred years ago.[7] It is nothing to lose sleep over, unless you are a physicist or engineer who measures time by the millionth of a second and needs an absolutely consistent clock. Our planet's motion was too irregular to serve as an ultimate standard of time measurement. What could take its place? The ideal would be a periodic event in nature that could be measured with great precision, one that would never change.

In the 1930s Columbia University physicist Isidor Rabi went looking for a solution at the atomic level. Rabi realized that the electrons that orbit the nuclei of atoms shift from one energy state to another if hit by precisely the correct radio frequency. It was as if these electrons were a new kind of pendulum, swinging back and forth in sync with the radio waves. Unlike a standard pendulum, the rhythm of the electrons would be unaffected by friction, heat, gravity, or other physical forces.[8]

Work on an atomic clock paused during World War II, but by 1948 the National Bureau of Standards had built the first such device, based on the chemical compound ammonia. Impressive as this clock was, interactions between the hydrogen and nitrogen atoms in ammonia limited its precision. Scientists got better results with cesium, a soft, silvery metal like mercury. Cesium electrons change their state when hit by microwaves tuned exactly to 9,192,631,770 cycles per second, and they respond to the change by emitting a burst of energy at the same frequency. Thus, the atomic clock was born.

These clocks set a new standard for timekeeping. In 1967 the General Conference on Weights and Measures, a global standards-setting organization, decided to base the duration of one second on the frequency needed to shift the energy state of a cesium electron. Ever since, the world's official timekeeping organizations, such as America's National Bureau of Standards, have embraced the new standard, called Coordinated Universal Time.

Cesium clocks are extraordinarily accurate. The best of them will lose a second every 100 million years.[9] Because they are tuned to the superfast vibration of cesium atoms, the clocks can subdivide time down to the billionth of a second. This extreme precision makes the atomic clock the twentieth-century equivalent of John Harrison's pocket chronometer of 1761—a timepiece that transformed the art of navigation.

The clocks' pinpoint precision made it possible to measure the tiny variations in timing caused by distance. This capability was put to work in a navigational system called LORAN, which has been replaced by GPS in the United States and Canada, though it is still used in other countries, such as Russia. An upgraded version of a navigational system first used by the Allies in World War II, LORAN became even more accurate when atomic clocks were added.

LORAN uses multiple networks, or "chains," of radio transmitters. One serves as a master transmitter, while two to five others act as "slaves." Each master transmits a signal over and over, with a delay in each transmission that can be measured in microseconds. The delay is unique for every LORAN chain. For instance, the master transmitter serving boaters on the US Great Lakes had a delay of exactly 89,700 microseconds between transmissions.

When the master transmits a signal, the slave transmitters of the chain repeat it exactly a few microseconds later. A navigator with a LORAN receiver picks up these signals and times the microsecond-long delays between master and secondary broadcasts. Given the unique delay pattern for each series of signals, the receiver knows which LORAN chain it is listening to. The receiver can also measure the time it took for signals to arrive from the master and slave transmitters and uses them to calculate the distance between itself and each transmitter. The result is represented as a line on a chart. Performing the same calculation for each transmitter gives you multiple position lines, which cross at a single spot. And there you are.

Early LORAN users were required to do a fair amount of manual calculation; later computerized versions of the system would simply spit out latitude and longitude. The results were accurate enough for most navigational uses—between 0.1 and 2.5 nautical miles.[10] LORAN's accuracy depended upon its ultraprecise radio transmissions. If radio signals travel at roughly the speed of light, that is equivalent to 186.4 miles in a thousandth of a second. If the timing of a LORAN transmission is off by that margin, it becomes useless as a navigational aid. The signals needed to be timed by atomic clocks, capable of slicing each second into billions of precisely metered bits.

In thinking about updates to Transit, Easton realized that a space-based navigational system had to be as precise as LORAN. So he proposed building atomic clocks in his satellites. It had never been done before, but in the 1960s electrical engineers were working miracles of miniaturization as they sought to cram ever more scientific instruments into space satellites weighing only a few hundred pounds. Easton rightly foresaw that atomic clocks would soon be small enough to put into orbit.

Yet the US Navy and Easton were impatient; in their view they could learn a lot by experimenting with less precise timepieces. Until suitable atomic clocks were available, they used clocks that relied on the vibration of a piece of quartz. The forerunners of today's cheap quartz-crystal watches, these quartz timepieces were not nearly as accurate as atomic clocks, but were close enough for use in prototyping the technology.

In 1964 Easton and his colleagues at the Naval Research Laboratory began building a satellite called TIMATION I to test his space-navigation concept. It took three years to build and launch the satellite, but TIMATION quickly proved that Easton was on the right track. In tests conducted in October 1967, scientists pinpointed their location on Earth to within one-third of a nautical mile, simply by processing the radio time signal broadcast by the satellite. The system even worked when the receiver was mounted in a boat or airplane.[11] A second TIMATION satellite, orbited in 1969, proved less accurate because its electronics had been damaged by cosmic radiation. Still, Easton had shown that the basic concept would work. If the satellites were upgraded with atomic clocks, it would work even better, guiding aircraft or missiles to within a few feet of their targets.

Researchers on the air force's Project 621B team had reached similar conclusions independently. Between 1964 and 1966, engineers at Getting's Aerospace Corporation conducted a secret study of satellite-navigation concepts and concluded that a network of satellites with onboard atomic clocks was the path forward.[12] In 1969 the air force awarded contracts to four major defense contractors to begin designing a practical system.

Meanwhile, Getting made an unsuccessful bid to gain the backing of the White House. In February 1969, months before the first successful moon landing, the Nixon administration was looking for new missions for the civilian space program, preferably something that would provide another global demonstration of American scientific prowess. A panel chaired by Vice President Spiro Agnew was created to choose the next great goal in space exploration. Aerospace Corporation scientists proposed a global navigational system available for civilian use, with a separate ultraprecise mode reserved for the US military. The panel was unimpressed. "The Agnew Committee was looking for programs of greater propaganda value involving humans in space," wrote Getting later.[13] The National Aeronautics and Space Administration (NASA) committed instead to the building of the reusable space shuttle. To the White House, Project 621B was just numbers on paper.

Undeterred, between 1968 and 1971, the air force ran tests of the concept. Their "satellites" were actually balloons that beamed precise time signals to receivers on the ground below. Despite the primitive instrumentation, the researchers were able to calculate locations to an accuracy of fifty feet. As with the navy's TIMATION tests, the air force experiments left no doubt that the basic idea was sound.

Even the US Army had embraced the idea of measuring location from space. Between 1964 and 1969 the army launched a series of satellites called SECOR, for Sequential Correlation of Range. SECOR was intended not as a navigational aid for travelers, but rather as a tool to build better maps. SECOR would beam its radio signals to three receivers on earth whose positions were exactly known and to a fourth receiver placed in an area that had not been surveyed. The transmissions would enable geographers to calculate the exact location of the fourth receiver, relative to the other three. SECOR precisely located the positions of Pacific islands that had never been accurately mapped before.[14]

The army, navy, and air force were each spending millions on their own space-based navigational systems. To Pentagon planners this seemed an absurd waste of resources, especially in the midst of the expensive war in Vietnam. It was obvious that all armed services could benefit from a single system, available to soldiers, sailors, and airmen alike. However, no one at the Defense Department seemed ready to settle on a unified approach.

That began to change in November 1972, when Lieutenant General Kenneth Schultz, commander of the Air Force Space and Missile Systems Organization, went looking for a new leader for Project 621B. He settled on Bradford Parkinson, a colonel with twenty-six combat missions over Vietnam in AC-130 gunships. More important, Parkinson held a doctorate in astronautical engineering from Stanford and had chaired the Astronautics Department at the US Air Force Academy.

Parkinson recruited a team of first-rate engineers. Just as important, he won the ear of Malcolm Currie, director of defense research and engineering at the Pentagon and one of the most influential men at the Defense Department. After an in-depth briefing from Parkinson in the spring of 1973, Currie understood the immense military potential of satellite-based navigation. Although a review board chaired by Currie and composed of officers from all services voted to halt the air force program, the project was far from dead. Currie ordered Parkinson to start over, this time by devising a system that could earn the support of all the armed services.

Work on the new venture began on Labor Day weekend of 1973, and despite Currie's admonition to develop a joint program, the air force was the only branch represented at the meeting. Still, Parkinson had been consulting with navy scientists ever since he took over the air force project, and his group was happy to incorporate the most attractive features of the navy's TIMATION technology into their system.

At the top of the list was building atomic clocks into the satellites. In 1972 a German company called Efratom Elektronik delivered the first atomic clocks small enough and resilient enough to launch into orbit. The Naval Research Laboratory would now build two of the clocks into its next TIMATION satellite. In addition, the Labor Day committee opted for an orbital configuration similar to that proposed by Easton. A network of twenty-four satellites would whirl constantly overhead, each of them completing an orbit in eleven hours and fifty-eight minutes. They would be launched into four circular orbits, each tilted 63 degrees relative to the equator, with eight satellites in each orbit. Each satellite would soar far to the north and south as it circled the planet. Beneath them, the earth would rotate east and west on its axis. In effect, the orbiting satellite network would be synchronized with the spinning planet, so that there would always be several satellites overhead virtually anywhere on earth. The Easton plan would later be modified for better coverage, with four satellites placed in six orbital paths, but the principle would remain the same.

Incorporating so much TIMATION technology helped bring the navy onboard. And there was even a part for the army. Its vast Yuma Proving Ground in southwestern Arizona would be used for ground-based testing of the system.[15] With a single system that would satisfy the army, navy, and air force, the Defense Department was finally ready to get on board. In May 1974 it committed to the new NAVSTAR Global Positioning System. Despite appearances, NAVSTAR wasn't an acronym; someone in the Pentagon just liked the sound of it.

The first two NAVSTAR satellites were in fact the last two in the TIMATION series. TIMATION III, now renamed Navigation Technology Satellite (NTS) I, went aloft in July 1974. Its two pairs of rubidium clocks were the first atomic clocks ever put into orbit. It was not a success: glitches in the satellite's attitude control system exposed the satellite to excessive sunlight, which caused overheating problems in the clocks.

Three years later Navigation Technology Satellite II, the last of the TIMATION family, flew into orbit with a pair of cesium atomic clocks, the most accurate timepieces flown so far. NTS II was arguably the first true GPS satellite, though it was far from capable of real-world use. Instead, its precise radio signals provided the essential data needed to build an operational system.

Among the most crucial observations was confirmation of one of Albert Einstein's odder predictions. Einstein's theories of general and special relativity held that although time might seem relentless and inexorable to humans, it is actually quite variable. For instance, time passes more slowly for someone on a moving train than it does for a person sitting on a park bench, watching the train roll by. Moreover, Einstein argued that time would move faster for a person floating in space than it would for a person on earth, because as you draw closer to a planet's gravity field, time slows down.

These effects are so tiny that an unaided human would never notice them. An atomic clock, however, capable of measuring time down to the nanosecond, might pick them up. Sure enough, the cesium clocks on board NTS II ran a little slower as they sped around the planet at more than seventeen thousand miles per hour. Yet the same clocks were sped up an even greater amount, because of their altitude above the planet. In all, the time signals broadcast by these highly accurate, highly predictable clocks were too fast, by about thirty-eight microseconds per day. Once again, Einstein had been proven right.

This was not merely a quirky theoretical insight. It posed a critical challenge in building a GPS system. As mentioned earlier, GPS receivers use a transmitted time signal from the satellite to calculate position. If that time is off, the resulting location fix can be off by miles. The user must have signals from at least four satellites to get a solid location fix, but if all four of them are transmitting the wrong time, the system would cease to function.[16]

The NTS II experiment left no doubt that GPS would have to compensate for the Einstein effect. And it does. Before each satellite is launched, its clocks are set to compensate for the relativistic effects of high altitude and a fast orbit. This adjustment greatly reduces any errors, but small variations in each satellite's orbit can cause timing variations. Complicating matters, the small orbiting clocks have minor imperfections that can introduce random errors. So the entire network of satellites must be updated once a day with a time signal beamed from a superaccurate atomic clock on earth.

These daily updates tell each satellite its exact position over the planet. An earthbound network of radars helps pin down this location. In addition, the satellites' signals are constantly being collected by ground stations scattered around the globe. Using methods similar to those employed by the scientists at Johns Hopkins who tracked Sputnik in 1957, the US Air Force always knows the position in space of each satellite. Every day this location data, and a clock correction, is beamed to every satellite in the GPS constellation.

Given the innate complexity of the GPS system, its relatively sparse budget—$150 million to start—and the fact that this program lacked the self-evident drama of the Apollo program, it is perhaps not surprising that the completion of the GPS network took about twice as long as putting a man on the moon.

For some people, that was way too long. In September 1978, in the sky over San Diego, a tiny Cessna 172 aircraft blundered into the path of a Boeing 727 jetliner. The resulting collision killed everybody on board the two planes, along with 7 people in the city below—a total of 144 dead. It was unmitigated horror, which Gerald O'Neill believed might have been prevented had the aircraft been equipped with better navigational aids. Like the developers of GPS, O'Neill looked to the sky. An avid pilot, O'Neill was a former professor of physics at Princeton University and a fervent advocate for human colonization of space (he gained national fame for his book on the topic, titled *The High Frontier*). O'Neill thought he could construct a relatively cheap and simple space-based navigational system. He proposed installing special transponders on three communications satellites parked in a stationary orbit over the United States. Subscribers to the service would carry devices that could transmit a signal to the satellites. The satellites would relay the signal to a ground station crammed with supercomputers, which would calculate the subscriber's location, based on the time at which each satellite received the signal, and relay the location to the subscriber.

O'Neill patented his idea and turned it into a business: Geostar Corporation. He raised millions from investors to license the necessary radio frequencies and to piggyback his radio transponders onto communications satellites. By 1988 thousands of Geostar tracking units were in use, many by trucking companies that used the system to pinpoint the location of valuable loads. The early Geostar receivers cost as much as four thousand dollars, but, like other electronic devices, became cheaper over time. O'Neill figured that, in time, Geostar location detectors would become as common on airplanes as altimeters and their precise measurement of a plane's location would prevent future tragedies like the one over San Diego.[17]

Alas for O'Neill, his company ran headfirst into a potent rival, a then-obscure San Diego startup called Qualcomm, which developed a more sophisticated system called OmniTracs. By 1991 Geostar filed for bankruptcy; the following year O'Neill died of leukemia. Qualcomm went on to become the world's dominant maker of cell phone chipsets. Along the way, it became a leader in putting GPS technology into millions of hip pockets.

First, however, the United States had to build its GPS satellite network. Between 1978 and 1985, the air force lofted ten first-generation satellites into orbit. These satellites, called "Block 1," were launched solely to enable engineers to test the hardware and software needed to make the system work. Yet all was not smooth sailing. In May 1981 the Armed Services Committee of the US House of Representatives voted to cut the program altogether, fearing it would prove too expensive. Later the money was restored to the budget. In 1981 one of the early satellites was destroyed during an abortive launch.

The Pentagon was determined to get full value for its money. Along with GPS equipment, the later Block 1 satellites carried sensors capable of spotting nuclear detonations from space and radios to relay the information to Washington. GPS satellites were ideal for the task, because their orbital path let them oversee so much of the planet's surface. These "NUDET" sensors are still standard equipment on today's GPS satellites.

The air force expected to begin launching its more capable Block II satellites in 1987. But on January 28, 1986, the space shuttle *Challenger* detonated in the sky over Cape Canaveral, killing seven astronauts and crippling the American space program. The shuttle had been the official launch vehicle for all GPS payloads. The disaster, caused by the cracking of a cheap piece of rubber in a solid rocket booster, revealed the fragility of the shuttles and the risk to national security in relying on them for satellite launches. The air force reoriented the program, opting instead to use unmanned Delta rocket boosters. The shift in strategy added two years of delay, and there would be no GPS satellite launches until 1989.[18]

Despite such setbacks, even in its early stages the GPS network enabled remarkable feats of engineering. In May 1983 a private jet owned by defense contractor Rockwell International flew from Cedar Rapids, Iowa, to Paris, becoming the first airplane to navigate across the Atlantic guided solely by GPS. The incomplete GPS network provided only limited coverage, so the aircraft landed in Vermont, Newfoundland, Iceland, and the United Kingdom when the satellites passed out of range. As a result, the trip took four days. But when the Sabreliner jet parked at Le Bourget airport—the same field where Charles

Lindbergh had set down after flying the Atlantic alone in 1927—its nosewheel rested within twenty feet of its intended parking space.[19]

It is notable that this first Atlantic crossing was made by a private jet. Even though GPS was a military program, it had always been intended for civilian use. "Contrary to some versions of GPS history, from the very beginning, GPS was configured to be a dual-use system," wrote codevelopers Bradford Parkinson and Stephen Powers.[20]

Some lost sight of this fact in the aftermath of a Cold War tragedy. On September 1, 1983, a Korean Air Lines 747 departed Anchorage, Alaska, en route to Seoul. It seemed inconceivable that KAL Flight 007, flown by two of the airline's most experienced pilots, could somehow lose its way. Yet this plane blundered into the airspace of the Soviet Union, where it was promptly shot from the sky by a fighter plane. All 269 passengers and crew perished.

It was one of the ugliest moments of the Cold War. Furious US officials accused the Soviets of cold-blooded murder; the outraged Russians insisted the plane had been on a spy mission.[21] In the immediate aftermath, President Ronald Reagan announced to the world that to prevent future tragedies, the GPS technology being developed for the military would be made available for civilian use as soon as the satellite network was made operational.

We will never know whether having GPS on board would have saved KAL 007. The plane's navigational systems were superb. There was Doppler radar, capable of identifying the aircraft's location by bouncing radio signals off the earth below. As Doppler began to be installed on planes in the early 1960s, airlines that had used full-time navigators on international flights began to abandon the practice; the pilot and first officer could find their way around the world without the extra help. Then there was inertial navigation. The Boeing 747 had been the first commercial airliner to use an inertial navigator; indeed, each plane carried three of them, for extra reliability. The system had proven extremely reliable and quite accurate, but only when used properly. An investigation by the International Civil Aviation Organization concluded that KAL 007 went wrong because its flight crew blundered while setting up the system. The inertial navigator should have steered the plane to a series of preprogrammed waypoints in the air, each point located well outside Soviet airspace. Merely by turning one knob to the wrong position, the navigation unit ignored the waypoints and instead followed a magnetic compass heading that took the doomed jet deep into the danger zone. Had the flight crew paid closer attention to their instruments, or used the Doppler system to double-check their location, they would have likely caught the error in time. But they didn't. If they had had a GPS system, they may have ignored it as well.

Causes of the crash aside, Reagan's seemingly magnanimous offer of GPS technology was a clever propaganda coup, no more. The GPS satellite network had always been designed for civilian use and would have been made available to commercial navigators in any case. Indeed, a notice posted two years earlier in the *Federal Register* had said that the completed network "will be made available to the worldwide civil/commercial community within the limits of national security considerations."[22]

The military was wary of their earlier generosity, however. The more the generals began to appreciate the possibilities of GPS, the more they worried about the risk of providing everyone on earth with a means of superaccurate navigation. Providing public access to the navy's Transit system was harmless enough. That system could give a navigator an accurate fix on his position, but it could not provide constant updates as he continued his voyage. He would have to rely on less precise navigational aids between satellite fixes.

With GPS satellites, by contrast, a user could receive second-by-second position updates. A self-guided bomb or cruise missile, fed a stream of GPS data, could constantly adjust its aim as it flew, down to the moment of impact. Given the exact latitude and longitude of a target, a bomb could land within a few feet of dead center. Armed with such precision-guided weapons, American forces could accomplish their missions with fewer aircraft and fewer bombs. Tom Logsdon, an engineer who spent twenty years working on GPS, calculated that because of its greater accuracy, a single GPS-guided weapon would be as effective as five to fifteen unguided bombs.[23]

But if anyone could access the GPS system, foreign enemies could also build guided weapons and use America's own satellite guidance system to lob them at American targets. The Pentagon attacked the problem by designing GPS as two navigational systems in one. Its military channel, encrypted and inaccessible to civilian users, would guide weapons and troops with pinpoint precision. A separate civilian channel would be far less accurate, but good enough for everyday use by ship captains or truck drivers.

Each GPS satellite broadcasts a constant stream of signals over two different radio frequencies. One signal, called the "coarse acquisition" (C/A) code, was intended for civilian use and was broadcast over one of the two frequencies. The other was called the "precise" code and was meant for the military. The P code delivered more accurate navigational results because it was broadcast over both radio frequencies simultaneously.

Back when William Guier and George Weiffenbach learned to track radio signals from Sputnik, they discovered that the ionosphere, a layer of the atmosphere full of charged subatomic particles, tended to distort the incoming signals. However, when the Russians launched Sputnik II, they put two radios on board. The team at Johns Hopkins soon found that by comparing the two signals, they could clean up the atmospheric distortion and get a sharper fix on the satellite. The P code system works the same problem in reverse. Military GPS receivers pick up both P code signals and automatically adjust for the ionospheric effect. As a result, they were expected to generate significantly more accurate location fixes than civilians would get with their C/A code receivers. Military-grade GPS receivers were expected to enable accurate navigation to within 20 meters of the intended location; that's about 66 feet.[24]

Even so, the designers of civilian GPS had done their work too well. Researchers found that despite atmospheric distortion, a civilian GPS system would still be accurate to within 20 or 30 meters. So the designers of GPS decided to cheat. In 1989 when the program began in earnest with the launch of the first Block II satellites, which would form the backbone of a fully operational network, the satellites featured a cunning bit of self-sabotage. Errors could be deliberately added to the C/A signal, misstating the time on each satellite's atomic clock and the satellites' precise location in space. Thanks to this gimmick, known as "selective availability," a civilian GPS device would be accurate to within only 100 meters, or 330 feet—about the length of a football field.[25]

This is adequate for a variety of civilian uses. Get a sailboat to within 300 feet of home port, and the skipper can probably eyeball it the rest of the way—if the weather is clear and sunny. If the sky is dark and the weather is dirty, however, and you're piloting a fifty-thousand-ton freighter through Sandy Hook Channel into the harbor of New York, being out of position by 300 feet is not an option.

It certainly was not acceptable to the United States Coast Guard. Charged with securing safe passage through the nation's ports, this federal agency wanted civilian GPS to be as precise as the military version. So the coast guard found a way to defeat selective availability—a system called differential GPS.

The agency built dozens of radio transmitters at fixed, precisely mapped locations, covering the nation's seaports and inland waterways like the Great Lakes and Mississippi River. Each transmitter broadcast its location, as well as a signal that corrected the erroneous data coming from the satellites. Differential GPS receivers combined supplemental broadcasts with standard GPS to produce location fixes that were far more accurate than civilian GPS; indeed, they were often better than military GPS, nailing down the user's location to within 30 feet. The air force was not happy about such interference, but the coast guard had the backing of shipping companies and the US Congress. Besides, the coast guard showed the futility of trying to dumb down GPS. A determined enemy would find a way past selective availability. "The cat's out of the bag," air force major Steve Gabriel told a journalist in 1993. "If you can build a missile, you can probably figure out how to do this."[26]

The coast guard was not the only branch of the US government looking to improve on the precision of civilian GPS. The idea of hyperaccurate satellite navigation also appealed to the Federal Aviation Administration (FAA) as the ideal tool for landing airplanes in bad weather. Major airports throughout the world use the Instrument Landing System, or ILS, a costly and sophisticated service descended from the land-based radio beacons described in earlier chapters. A pilot trained for instrument landings can establish his aircraft's proper direction and rate of descent by watching onboard gauges or a video monitor, enabling safe landings even in darkness and bad weather. Yet hundreds of small airports lack ILS systems, increasing the risks to operators of small aircraft. A GPS system might serve as an ILS substitute, if it could show a pilot his plane's altitude and location to within a few feet. Of course, civilian GPS was not nearly accurate enough; for that matter, neither was the military version. However, as the coast guard had surmised, the system's accuracy could be dramatically improved with a secondary radio signal that transmitted corrections to the civilian system.

The FAA's version, called the Wide Area Augmentation System, or WAAS, used ground stations that picked up the GPS signals, corrected their inaccuracies, then sent this corrected data to communications satellites in geostationary orbits over the United States. Airplanes would use navigation radios that combined incoming GPS signals with the WAAS satellite data to deliver location fixes accurate to within about six feet.[27] Developed in the mid-1990s, the service has since become popular with amateur pilots and operators of private jets. In late 2009 Horizon Air, a regional carrier, became the first commercial airline to use WAAS navigation; in 2012 the company announced that it would install the technology on its entire fleet of aircraft.

Even the US Army found itself chafing under the limitations imposed on civilian GPS accuracy during the 1990–1991 Gulf War. By that time only sixteen GPS satellites were in orbit, but already they were proving vital tools for guiding infantry troops through the deserts of Iraq. The trouble was that there were not enough military-grade

GPS units to go around. Desperate to get the technology to the troops, the US military bought thousands of civilian GPS units, which at the time were mostly being used by backpackers, hunters, and land surveyors. Thanks to selective availability, these units were not nearly as accurate as the military kind. The Pentagon finally ordered that selective availability be switched off, and in August 1990 it was. Instantly, civilian GPS units became almost as accurate as their military cousins. It was too good to last; with Saddam Hussein safely defeated by July 1991, selective availability was reactivated and civilian GPS restored to its previous level of mediocrity.[28] By December 1993 a full complement of twenty-four GPS satellites was in place, providing worldwide service to civilian users. In April 1995 the network's secure military version was declared fully operational.

In the nineties GPS was still science fiction for most consumers and of limited value to potential commercial users. A hyperaccurate version of the technology would be a godsend for land surveyors, trucking companies, even farmers who could use GPS-guided tractors to precisely plant and harvest their land. And although car navigation systems based on GPS were being developed by the mid-1990s,[29] the Pentagon's selective-availability mandate ensured that they were not nearly good enough to provide the turn-by-turn driving directions we now take for granted.

Businesses that stood to profit from the technology chafed against the constraints of the Pentagon policy. Meanwhile, the Russians were developing their own version of the technology, which they called GLONASS. Their network of twenty-four satellites, which went live in 1995, had much in common with GPS. GLONASS could be used by anyone in the world; each satellite broadcast both a military and a civilian signal, and the civilian signal was surprisingly accurate, capable of pinpointing the user's location to within about sixty feet. There was one big difference—the Russians never bothered with selective availability. GLONASS could give civilian users a much-sharper location fix than the deliberately fudged civilian GPS ever could.[30]

Thanks to post-Soviet economic turmoil, the GLONASS system fell apart for lack of maintenance. It was eventually revived after the turn of the century under Vladimir Putin. Today, GLONASS is fully operational, and apart from being a valuable service in its own right, the Russian network has become a useful backup to GPS. A number of devices, such as the Apple iPhone 4S and iPhone 5, can get a fix on both GPS and GLONASS satellite signals, generating more accurate location fixes than they would obtain from a single network. Still, the decay of GLONASS during the 1990s prevented it from becoming a major rival to GPS, despite its distortion-free civilian service.

The nations of Europe offered yet another challenge to the dominance of GPS. Unwilling to be entirely dependent on the United States for satellite navigation services, in early 1999 the European Union announced that it would create a new network called Galileo. Although it would have a military component suitable for use by the EU's armies, navies, and air forces, Galileo would be designed primarily for use by civilians and businesses.

Galileo would offer a two-tier system—a free service capable of measuring the user's location to within about three feet and a commercial service for businesses that needed hyperaccurate location data, such as land surveyors and mining companies. These users would pay for access, but in exchange they would be able to measure locations to within a fraction of an inch. Neither system would include selective availability, like the American GPS network. Indeed, the decision to build Galileo was driven in large part by European fears that America might degrade GPS in times of military conflict or shut down civilian access altogether. French president Jacques Chirac warned that without a satellite system of its own, Europe would become little more than a vassal of the United States.[31]

Eager to retain its dominance in space-based navigation, the United States lobbied European leaders to drop the idea of Galileo, but to no avail. America had to settle for persuading the Europeans to modify their system to ensure it would be interoperable with GPS and also that it would use a military frequency that would not interfere with GPS transmissions.

By 1996 the United States decided that selective availability was more trouble than it was worth. In 1996 the Clinton administration announced it would shut down selective availability at an unspecified future date. Concern about competition from GLONASS and Galileo was one reason that the switch was finally thrown on May 1, 2000.[32] The United States further undercut Galileo by deciding not to charge user fees for access to civilian GPS services. Whereas the Europeans would bill customers for their most precise location services, the entire cost of GPS would be borne solely by US taxpayers.

Around the world many commercial GPS users feared that the United States would tamper with the accuracy of the service during some future conflict. Yet even after the terrorist attacks of September 11, 2001, and the subsequent wars in Iraq and Afghanistan, selective availability was not reactivated. In 2007 the United States officially pledged that selective availability would never be used again. In fact, in 2014 it will begin launching a new generation of GPS satellites that lack the capacity for selective capability, because the necessary hardware will no longer be installed.[33]

Eliminating selective availability was not just good news for navigators. It was a boon to the world's ability to tell time. The time displayed on your cell phone's screen is extremely accurate and constantly updated through the phone's radio link to the nearest cell site. It was not designed this way so you could stop wearing a wristwatch. As you travel, the cellular system has to switch calls from one cell to the next without dropping the connection. It does not always work, but it would not work at all unless the phone network could synchronize the relay of digital radio packets down to thousandths of a second. In other words, the digital clock in every cell must be synched to a highly precise master clock. The same is true of landline phone services that move hundreds of simultaneous calls over a single fiber-optic thread. Each call is a stream of laser pulses that must be precisely synchronized with all the other pulses on the line to ensure that every call gets through.

Phones are not the only systems that rely on a hyperprecise time standard. The alternating current used in the US electrical system pulses at a precise sixty cycles per second. If a power plant's generator is off by even a fraction of a second, the variation can damage sensitive electronic devices in homes, offices, and factories. Banks and financial institutions handling millions of transactions a day also use high-precision time standards to stamp every purchase and sale.

The US Navy began transmitting precise time data via telegraph in 1903; the National Bureau of Standards started doing it by radio in 1923. But as was the case with tracking Sputnik, variations in the ionosphere introduced tiny but significant inaccuracies into the signal. Early in the space age, scientists realized that transmitting the time signal from orbiting satellites would enable far more accurate results.[34]

GPS was exactly the sort of system they had in mind. Its own operation required constant transmission of the exact time to millions of GPS receivers on the ground. The system's control center constantly updated the satellites' atomic clocks, ensuring their accuracy. Around the world, people could simply tune in to the satellites as they passed overhead. This could not be done in the era of selective availability, but switching off this feature restored the integrity of GPS time signals, allowing time-sensitive enterprises worldwide to set their clocks by GPS. Today these orbiting satellites are the world's primary precision timing standard.

At last, the civilian GPS service was accurate enough to be useful to timekeepers and navigators alike. It was good news for backpackers, hunters, and land surveyors who had grown to love their handheld GPS location trackers. By 2000 a decent unit could be purchased for a few hundred dollars. Still, not many people felt a need to know their latitude and longitude at any given moment, which was why only about 4 million civilians worldwide owned a GPS device. It was a decent number, but hardly a mass market.

That same year nearly 110 million Americans owned a cell phone, a number that had increased tenfold in eight years. The construction of nationwide cellular networks and the falling prices of the phones made them intensely attractive to consumers. However, cell phones created an unexpected problem for law enforcement and emergency services—the same one that had gotten Karen Nelson into trouble.

As millions became dependent on mobile phones, locating a caller in distress was becoming a lot harder. A research study found that 25 percent of 911 callers in Los Angeles were too badly injured to give their location or simply did not know where they were.[35] There were countless stories in which cell phones saved lives, but a significant number of grim anecdotes about callers who could not be located until it was too late.

Regulators at the Federal Communications Commission (FCC) could see where things were heading. Tens of millions of Americans were becoming dependent on phones that were not compatible with Enhanced 911 (which showed the street address associated with a phone line), and unless that changed, desperate people were going to die. Thus, in 1994, when only about 24 million Americans had cell phones, the FCC declared that the nation's wireless carriers must find ways to deliver Enhanced 911 services. That meant that when a call arrived from a cell phone, it would have to include an approximate fix on the physical location of the phone and consequently the person using it.

It was a massive technical challenge, but far from impossible. The phone company could triangulate the signals from several cell towers and calculate the approximate location of a phone, based on the time when the cell phone signal reaches each tower. Even a difference of a few thousandths of a second between the signal's arrival at towers A, B, and C enables location of a phone to within about 150 feet.[36] This method, called time difference of arrival (TDOA), was exactly the method chosen by major US cell phone companies like AT&T and T-Mobile USA. It is an attractive approach partly because it allows customers to continue using their existing phones. All the necessary hardware and software is installed in the cell towers.

Other engineers looked to the multibillion-dollar GPS satellite system. A phone with an integrated GPS receiver could be tracked down with a degree of accuracy that the TDOA system could not match. Such an idea drove Robert Tendler, a Boston patent attorney, to try to market FoneFinder in 1997. FoneFinder was a standard low-cost cell phone with a GPS device bolted onto it, featuring a red emergency button. When this button was pushed, the phone would dial 911, while the GPS unit would determine the user's location. The GPS had a speech synthesizer that would announce the latitude and longitude in a loud electronic voice, so that the emergency dispatcher on the other end could jot it down. He could then type the numbers into a mapping database that would look up the correct street address.[37]

FoneFinder was one of the earliest attempts to meld GPS and cellular technology. It earned a place in the Smithsonian Institution as a valuable historical artifact. As a commercial product, however, it flopped.

The man who perfected phone-based GPS technology, Stephen Poizner, is not especially famous as an inventor. Later in life, Poizner went into politics, was elected California's commissioner of insurance, and then lost a bid to become governor of the state. Poizner's previous career as a high-tech entrepreneur was mostly forgotten, even though SnapTrack, the company he founded, was arguably as important to modern navigation as Sperry Gyroscope or Draper Labs. It was Snap-Track that figured out how to squeeze a full-fledged GPS system into a phone, an achievement that has permanently altered our sense of place.

By Poizner's account, it was not an easy sell. "Back in 1995, it was really difficult to raise venture capital for the idea of putting GPS in a cell phone," recalled Poizner. "It was viewed with massive skepticism." Cell phone makers wanted no part of a system that would add significantly to the cost of a handset. SnapTrack's engineers would have to put all the GPS electronics onto a single chip or, better yet, a set of circuits that would fit on an existing chip.

Poizner attributes his success to hiring a talented team of scientists and engineers, among them chief technology officer Norm Krasner, an expert in dealing with Code Division Multiple Access radio signals, the technical standard used by GPS satellites to broadcast their location and timing data. These CDMA signals were weak, and developing chips and antennae that could detect them and fit into a low-powered phone required some near-miraculous engineering.[38]

SnapTrack augmented the feeble GPS signals with help from the cellular network, in a hybrid approach called Assisted GPS. A "pure" GPS system relies entirely on the signals from orbiting satellites, but acquiring the signals from four satellites can take several minutes. In Assisted GPS, the cellular network transmits vital time and location data to the phone, giving it a head start, and letting the phone calculate its location in seconds.

SnapTrack had to overcome an additional problem: the FCC. When the FCC announced its original Enhanced 911 standards in 1996, it insisted that phone networks be able to locate the position of older phones. This was no problem for TDOA location systems, in which the technology was stored at the cell towers. A GPS system, however, would require everyone to buy a new phone. The FCC would later relent, after realizing that most cell phone users traded in their handsets every couple of years anyway.

Now cell phone makers began taking SnapTrack seriously. The company licensed its technology to chip makers like Motorola and Texas Instruments. The turning point came in 2000, when the giant Japanese cell phone carrier NTT DoCoMo became the first in the world to sell GPS-enabled phones using SnapTrack technology. Within weeks SnapTrack was purchased by the giant cell phone chip maker Qualcomm for $1 billion.

Qualcomm designed phones based on CDMA technology, the same radio signaling method used by GPS satellites. Moreover, SnapTrack's engineers were also CDMA specialists. Phones based on CDMA soon took the lead in GPS integration, as Qualcomm built SnapTrack's technology directly into its standard phone chipsets. Two of the four biggest US carriers, AT&T and T-Mobile USA, used the Global System for Mobile Communications, or GSM, a technology popular in most of the world. GSM could be integrated with GPS, with a little extra engineering effort. But the other two American giants, Verizon Wireless and Sprint, ran CDMA-based networks; these companies easily and quickly adopted Qualcomm's SnapTrack system. By 2001 Verizon Wireless and Sprint were selling thousands of phones that included GPS support.

These changes were not immediately obvious to the average user. Early GPS phones used the technology solely to direct first responders to the caller's location in an emergency. They did not include software applications to allow a phone user to track his location.

Soon enough, companies like Irvine, California–based Networks in Motion developed consumer-friendly navigation programs like Atlas-Book that ran even on the cheapest low-powered phones. When a user typed a destination into the phone keypad, it was relayed to a remote computer that calculated the best route to the address. Using the cell phone's data connection, AtlasBook sent back an on-screen route map and turn-by-turn driving instructions, uttered in a reasonably human-like voice.

By 2004 Americans had purchased 25 million GPS-capable phones, mostly through Verizon Wireless and Sprint. Most were of the simplest sort, flip phones that came free with a two-year contract that people used for placing calls. These phones had a limited capacity to run software apps, but they could handle simple navigation programs like AtlasBook, which Verizon Wireless marketed under the name VZ Navigator.

The smartphone reset everyone's expectations for portable navigation. Consumers grew accustomed to treating their phones as pocket computers. Even though AT&T and T-Mobile had rejected GPS as the best way to meet the FCC's 911 standards, the companies began adding it to their phones as a must-have feature for bewildered travelers. AT&T released the original iPhone, the device that truly launched the smartphone era. Ironically, the first iPhone came under fire for lacking a GPS chip; Apple tried to recoup by incorporating a technology that used the signals from local Wi-Fi Internet routers to pin down the phone's location. It amounted to a shamefaced concession that location and navigation services had become essential features in a state-of-the-art device.

All future iPhones were GPS capable, as were other rival handsets such as Research in Motion's BlackBerry phones. Google insisted on GPS in all phones that would run its Android operating system, launched in 2008. A year later Google built turn-by-turn navigation into Android as a free feature, eliminating the need to download a third-party app and pay a monthly fee. In 2012 Apple finally climbed onto the bandwagon by adding free navigation to its updated iPhone software.

According to market research firm ABI Research, Americans bought 228 million cell phones in 2011; of these 211 million were GPS capable—more than 90 percent. Worldwide, 1.6 billion phones were sold, with GPS built into 619 million of them, or almost 40 percent.

The success of GPS represents a remarkable return on investment. According to Anthony Russo, who oversees space-based navigation systems at the US Department of Transportation, since the 1970s, the US government has spent $19.6 billion to build GPS satellites and run the ground stations that keep the system operational. (That is not including the cost of launching the satellites into orbit, which may amount to billions more.) Still, GPS has proved its worth. One survey concluded that the technology now generates $96 billion in economic benefits to the United States each year. And that does not count its value to the US military or to the rest of the world.[39]

Who gets the credit for developing such a life-changing technology? That depends whom you ask. Now that GPS has become an indispensable tool of modern life, its surviving creators have engaged in a spirited dispute over which of them deserves more credit. On the one hand, air force engineer Bradford Parkinson cobbled together the joint program that got GPS off the ground. He was recognized for his work by the National Academy of Engineering, which in February 2003 awarded its Charles Stark Draper Prize—its most distinguished honor—to Parkinson and Aerospace Corporation president Ivan Getting for conceiving and creating the GPS system. The two men were honored again in 2004, when they were inducted into the National Inventors Hall of Fame. On the other hand, it was US Navy engineer Roger Easton who put many of the ideas into practice. Easton's TIMATION satellites applied the concepts that would eventually become key features of GPS, like the inclusion of space-based atomic clocks.

Many people think that Easton deserves a share of the credit for GPS. Easton himself thinks he should have gotten the lion's share of it. "I can't think of any ways that GPS improved on Timation," Easton said in 2008. "Essentially, they are the same system."[40] The Bush administration acknowledged that Easton had a point. In 2004 Easton was awarded the National Medal of Technology, the US government's highest award for invention, for his role in developing GPS. In 2010 he joined Getting and Parkinson in the National Inventors Hall of Fame.

Parkinson insists that Easton's contributions have been overrated. "For all the years I ran the program, I knew Roger was claiming it was all his idea," he said in a recent e-mail message. "My people were incensed and wanted another fight. I said no—the important thing was getting it built and I had my hands full with running the program and defending the budget in the Pentagon." Still, Parkinson couldn't resist adding, "Roger had nothing to do with the actual developments."[41]

Of course, both Parkinson and Easton made crucial contributions. Easton certainly was the first to put atomic clocks into space. And Parkinson's adept leadership of the project overcame rivalry and infighting between the armed services, as well as an apathetic Congress that came close to canceling the entire project.

BETWEEN THEM, EASTON AND PARKINSON HAD DEVELOPED A TECHNOLOGY that would make navigators of us all. However, the task of putting the world in our pockets was not yet complete. On its own a GPS system merely spits out latitude and longitude data. To humans the world is a complex image made up of rivers and forests, streets and skyscrapers. Some of us describe the world in numbers, but all of us see the world in pictures. GPS was a new kind of compass, but to make the most of it, we would need a new kind of map.

The Accidental Navigators

LONG BEFORE PROCTER & GAMBLE USED RADIO TO SELL SOAP, ENGINEERS were using it to help travelers find their way. As we have seen, governments and corporations made vast investments in radio navigational networks and the space-based GPS radios that have begun to supplant them. A century later, new radio navigation system emerged quite by accident. Like the old one, it cost billions to build. However, the investment was made a hundred dollars or so at a time by millions of ordinary people—anyone who has ever visited the local electronics store to buy a Wi-Fi Internet router.

A small Boston company called Skyhook Wireless came up with what was probably the biggest advance in terrestrial navigation since the birth of GPS. It figured out a way to navigate much of the planet by tuning into those wireless router signals. Skyhook's Wi-Fi surveyors have created road maps of the world's major cities. A Wi-Fi-equipped laptop or smartphone, even one without GPS, can use Skyhook software to home in on its own location or to find the nearest hotel, hospital, or fast-food restaurant.

Skyhook works where GPS can't—in urban canyons or inside thick concrete walls that weak satellite signals cannot penetrate. Wi-Fi location is not a substitute for GPS; when the satellites are detectable, GPS gives a more precise fix. Instead, Skyhook markets a hybrid service that combines GPS and Wi-Fi data, offering whichever will give the most accurate result at a given moment. Skyhook works so well that Apple built the company's technology into the original iPhone, Intel is adding it to the company's laptop chips, and Sony is making it part of its handheld video-game devices. Google recognized the appeal of Wi-Fi location as well, but wanted to be in the driver's seat. The company launched its own version of the technology, using techniques that may have violated the privacy of millions around the world.

Yet despite what the controversy might suggest, the origins of this powerful technology are quite humble. Ted Morgan, the founder of Skyhook, and his business partner, Mike Shean, are neither navigators nor engineers. They are salesmen who salvaged the concept of Wi-Fi navigation from the rubble of a failed business venture, one that had nothing to do with helping travelers find their way. But then Wi-Fi itself was not conceived as an easy way to connect billions of people to the Internet. It began as a simple plan to improve upon the most unromantic of gadgets—the cash register.

Today you can walk into an Apple store, ask any clerk for an iPhone, and complete the purchase immediately. Your clerk can use an iPad tablet computer to ring up purchases and accept credit card payments anywhere in the store, because the tablet is connected via Wi-Fi to Apple's retail computer system. Apple's admirers rightly hail the system as a clever innovation. But the underlying concept was born in Dayton, Ohio, in 1986.[1]

NCR Corporation is based in Georgia now, and these days NCR is its official name. But those letters originally stood for National Cash Register. Founded in 1879, NCR long dominated the market for cash registers, massive, clattering mechanical contraptions used at supermarkets and department stores to process sales. I am old enough to remember them fondly, but if you are under forty, you have probably seen them only in black-and-white movies and *Wikipedia* photos.

Unlike most Rust Belt industrial companies, NCR did not sit idle as electronics rendered its core product line obsolete. In the 1950s the company became a major manufacturer of computers and later started production of the cash register's electronic heir, the "point-of-sale terminal"—basically, a digital cash register with a microprocessor at its heart. A point-of-service terminal could be easily programmed for the particular needs of any retailer, from a supermarket to an auto parts store. More important, POS terminals could be linked to a retailer's computer network, giving managers instant access to vital sales data.

POS terminals were great in theory, but NCR engineers chafed at the limitations of early computer networking gear. The company's retail customers had to invest in skeins of costly, cumbersome wire. Moreover, retailers often "refresh" their stores, changing the layout of shelves, aisles, and checkout stations. That meant running new network cables to the relocated POS terminals—another unwelcome expense.

But what if you could put the terminals anywhere you wanted? If you put a radio inside each POS terminal, it could transmit sales data to the central computer without wires. You could move the terminals at will, whenever the retailer wished to remodel. The NCR plan did not feature portable terminals hanging from belt loops, but it was a vital first step.

NCR soon found that creating a new wireless technology requires more than some software and a soldering iron. They also needed some real estate—the electronic kind. There is only so much electromagnetic spectrum to go around. The most useful frequencies run from 3,000 hertz, or cycles per second, up to 300 billion hertz. A variety of businesses and governments use these frequencies for everything from military communications to police radios to commercial radio stations. Which means not everyone can secure their desired frequencies.

The Federal Communications Commission was created by the US government in 1934 to allocate this limited radio bandwidth by issuing licenses for certain frequencies in certain parts of the United States. A license to broadcast can be extremely valuable. When the United

States switched from analog to digital TV broadcasting in 2009, the change freed up a vast chunk of frequencies. The federal government auctioned them off for use in cellular telephone networks. The sale brought in $19.6 billion.[2]

The radio spectrum was not always so highly valued. In fact, a few frequencies were set aside from the start as "garbage bands," allocated to industrial, scientific, and medical gear that worked by emitting radio waves. For instance, microwave ovens blast food with radio energy in the 2.4-gigahertz range. It was thought that if other broadcast services moved into the same frequency range, they would be overwhelmed by constant interference from millions of kitchens. Surely, no one would want such noisy, polluted frequencies.

It turned out someone did. In the late 1970s, an FCC engineer named Michael Marcus decided that the agency was letting good frequencies go to waste. He decided to bring to life an idea first suggested at the start of the twentieth century by an eccentric genius and a Hollywood starlet. The idea was to build radio devices that used more than one frequency. Transmitter and receiver would regularly "hop" from one frequency to another with the aid of a synchronizing system to ensure that both were on the same wavelength at the same time. The legendary Nikola Tesla won a patent on the idea in 1903.

Years later actress Hedy Lamarr secured a patent of her own. Under contract to Metro-Goldwyn-Mayer, Lamarr still found time to dabble in technical matters that had fascinated her since her abortive marriage to an Austrian arms merchant. She and a friend, film music composer George Antheil, designed a frequency-hopping system for radio-guided torpedoes to ensure they could not be defeated by enemy jamming signals.[3] By the 1950s frequency-hopping or "spread-spectrum" radios were being deployed by the US armed forces. The technology was treated as a military secret until 1976. Three years later there had been little effort to commercialize spread spectrum.

Marcus, however, realized that the technology could transform the FCC's "garbage band" frequencies into pure gold. Take the 2,400-megahertz band, the one used by microwave ovens. It actually contains dozens of separate frequencies. A radio that could rapidly hop between these frequencies would be effectively shielded from interference. A chunk of spectrum once thought worthless was now as valuable as any other.

The FCC could have made billions for the US Treasury by auctioning licenses for these bands. Marcus had a more radical idea—he proposed to give them away. Companies would be allowed to produce low-powered short-range radio devices broadcasting in these throwaway frequency bands, without the need to obtain federal licenses. Simply granting unfettered access to big slabs of radio spectrum outraged some FCC bureaucrats, who tried to force Marcus out of the agency. Yet his persistence won over the people who counted—the FCC commissioners with the final say. In 1985 they issued an order that essentially ratified the Marcus plan. Three radio frequency bands—at 900 megahertz, 2,400 megahertz, and 5,700 megahertz—were wide-open territory.[4]

NCR engineers were quick to recognize the value of the FCC's gift. They now had free radio spectrum to use in building their wireless data network. In 1987, at the company's research center in the Dutch city of Utrecht, engineers created a device that could transmit and receive data at ninety-seven thousand bits per second—not bad for a souped-up cash register. But there was better to come. A second-generation device delivered data at half a megabit per second, and it was clear that this new wireless technology had the potential to go faster yet.

In the spring of 1988, NCR was ready to start designing a commercial product—a circuit card that could be added not only to POS terminals, but to personal computers. However, the company dreaded a replay of the "standards war" then plaguing the users of wired computer networks. Two major technologies vied for dominance—a system called Token Ring, developed by IBM, and Ethernet, created by 3Com. Ethernet was destined to emerge as the global standard, used to this day, but for years the lack of a unitary standard made networking more complex and costly than it needed to be.

NCR had a better idea. It persuaded the Institute of Electrical and Electronic Engineers (IEEE) to develop a global standard for wireless computer networks. Devices built to that standard would work perfectly with each other, straight out of the box, no matter which company built them. Without waiting for the IEEE to complete its work, NCR introduced its first wireless networking product, WaveLAN, in 1990. The plug-in circuit cards cost $1,350 each, hardly affordable for a mass market. Yet as NCR had hoped, plenty of companies saw the benefits of linking their computers without wires.

Still, the technology was far from commonplace. That began to change in 1994, thanks to Carnegie Mellon University, a Pittsburgh school renowned for its computer science program. With funding from the National Science Foundation, the university purchased WaveLAN cards and wireless access points from telecom giant AT&T, which had acquired NCR for $7.4 billion in 1991. The result was called Wireless Andrew, in honor of Andrew Carnegie, the nineteenth-century steel magnate who founded the university. The network's data broadcasts linked seven campus buildings. For the first time hundreds of students and researchers could visit websites or search databases from a portable computer, without the need for a hardwired connection.[5] At first the high cost of WaveLAN gear ensured that only a few could use the network. By 1998, when Carnegie Mellon extended Wireless Andrew to cover the entire campus, cards for plugging into a laptop had fallen in price to around $350.

Wireless tech was about to become much cheaper, thanks to the IEEE. After seven years the engineering group finally settled on a standard for wireless data networking in late 1997. The new standard bore an inelegant name—IEEE 802.11. It made no difference to engineers and chip makers. With a universal standard in hand, any company could produce compatible wireless devices, ensuring a virtuous cycle of rapidly falling prices. That is, only if the technology caught on with millions of users. A better name would help, and in 1999 an alliance of wireless networking companies came up with one. The term *Wi-Fi* meant nothing in particular, but it had a catchy built-in rhyme that locked it into the brain. Besides, it hadn't already been copyrighted. Wi-Fi it was.

It was a good first step, but the new technology also needed a champion, a company that would back wireless networking with capital and prestige. It is not stretching the facts too far to say that the wireless Internet was born on a stage in New York City on July 21, 1999. Steve Jobs, chief executive of Apple, passed his company's newest laptop computer through a hula hoop while it downloaded Web pages live from the Internet. The thousands of guests at the Macworld trade show responded with roars of rapturous laughter. Message received: Look, Ma—no cables. Apple's new laptop computer would feature built-in Wi-Fi.

In Austin, Texas, Michael Dell was unhappy. The founder of Dell Computer, at the time the world's leading maker of personal computers, was well aware of the possibilities of wireless networking. He had wanted Dell to be the first to build it into laptops. Too late. Dell was not far behind. Neither were the other leading PC makers. Within weeks all of them offered Wi-Fi as an option.

So the virtuous cycle began. In 2005 electronics companies mass-produced more than 100 million Wi-Fi chipsets, according to the research firm ABI Research. A year later they had produced twice as many. The number jumped to 475 million in 2009 and exceeded 1.5 billion in 2012.[6] These days Wi-Fi chips are so cheap, they are featured in almost every intelligent device—video-game consoles, e-book readers, smartphones. To link up to these gadgets, consumers and businesses have installed hundreds of millions of Wi-Fi routers, access points to household or commercial data networks, and to the Internet.

What most people do not recognize is that every one of those access points is a navigational beacon. Match them up with latitude and longitude data, and a Wi-Fi device could become a sort of radio compass. It is the kind of idea that you would expect to come from the same people who invented GPS. Instead, it was generated by, in Ted Morgan's words, "knuckleheads from Boston with no scientific pedigree whatsoever."

Morgan and Shean were relatively successful executives at eDocs, a Massachusetts company that helped clients set up electronic billing systems. In 2003, a year before eDocs was acquired for $115 million, the two marketing executives decided to go their own way. Their new venture was inspired by their shared experiences on frequent sales trips. When they had a spare hour or two between appointments, both men would pass the time by firing up their laptops and connecting to a nearby Wi-Fi hot spot, operated by a business or broadcasting from someone's home. Morgan and Shean were struck by how easy it had become. "You could pretty much pull up into any parking lot, open up your laptop, find an open Wi-Fi signal and get your e-mail," Morgan recalled. "The more we did it, we didn't have to search as much. . . . [W]e were visibly seeing, month to month, the growth in Wi-Fi."[7]

It was like an engraved invitation to a couple of ambitious businessmen. Wi-Fi was becoming a billion-dollar business. Unlike other radio-based technologies, it was deregulated. Anybody with a good idea could forge ahead, without FCC approval.

At first, neither man knew exactly how they would profit from Wi-Fi. Then Morgan read about "mesh networking," a concept that seeks to overcome one of Wi-Fi's main drawbacks, its limited range. Wi-Fi devices generally have a range of a few hundred feet, enough to serve a home or a small business office. Yet some ambitious businesspeople and community activists wanted more. They saw Wi-Fi as a cheap way to deliver high-speed Internet access practically anywhere, bypassing the services offered by the big telephone and cable companies. Instead of spending millions to run copper wires or optical fibers, a company, nonprofit agency, or local government could set up a wireless network that would encompass the entire region at a fraction of the cost.

Of course, Wi-Fi signals lacked the range to cover a whole city or even an entire neighborhood. However, some engineers figured a workaround. A Wi-Fi router could be programmed to relay traffic from other nearby routers. In this way, multiple routers would form a mesh, like a strong but loosely woven net. Such a Wi-Fi network would function in much the same way as a cell phone system. A relatively small number of routers could cover a large area. Wi-Fi users in that area would never be out of touch. Wherever they went, they would stay connected to the network.

For a couple of years in the mid-2000s, Wi-Fi meshes seemed the best, cheapest way to deliver inexpensive Internet services in big cities. Community activists allied with vote-hungry politicians and ambitious technology companies to propose citywide Wi-Fi networks to serve Philadelphia, Houston, San Francisco, and Boston.

Morgan and Shean realized they might be able to sell the same technology to small Internet providers, who could not match the billions being spent by telephone and cable TV giants to upgrade their landline networks to support broadband data services. A wireless mesh network might deliver a competitive product at far less cost, if Morgan and Shean could develop the needed software. Neither man was an engineer. "I'm a product and marketing guy, and Mike's a sales guy," Morgan said. "We just got interested in it, immersed ourselves in it, taught ourselves it." In the end citywide Wi-Fi systems went nowhere, as designers found that building the networks was far more costly and complex than they had expected. As for Morgan and Shean, they spent six months trying to sell Internet service providers on the idea, without a single sale.

Yet their work had not been wasted. In order to demonstrate their concept, the two men had done a great deal of "wardriving"—cruising the streets in a car equipped with a laptop and software that detected nearby Wi-Fi hot spots. The wardriving software interfaced with a GPS sensor, revealing the wardriver's location and helping him plot the relative positions of the various hot spots. Their eureka moment came as Morgan and Shean played bocce and brainstormed ideas on the shores of a lake in Framingham. They'd been using GPS radio guidance to create a map of Wi-Fi signals. But once they'd completed this map, couldn't they dispense with GPS, and use the Wi-Fi signals for navigation?

The answer was surely yes. From the beginning, wardriver software had enabled hot-spot mapping, but only for the purpose of building a database of free Internet access points to help travelers needing a quick hookup. Morgan and Shean wanted to use the Wi-Fi signals as homing beacons.

Every Wi-Fi device, like every cell phone, broadcasts a unique digital identifier, called a Media Access Control (MAC) address. They look something like this: 00:16:B4:CF:61:1F. No two are alike. Multiply this by the millions of Wi-Fi routers in use around the world, and you've got the makings of a very accurate map.

You do not need the exact location of each Wi-Fi signal. Instead, the mapper drives a car equipped with a device that measures the strength of each radio signal and the direction it comes from. These factors would constantly change as the vehicle rolled down the road, as would the number of Wi-Fi signals detected. In sparsely populated places, the wardriving device would hear from a single router or none at all. In cities and towns, the antenna would often pick up multiple Wi-Fi signals of varying intensity and from different directions. The mapping system includes a high-quality GPS system that traces the exact location of the wardriving car, matching it to the strength and direction of incoming Wi-Fi signals, and of course the MAC address of each Wi-Fi router. The result, according to Morgan, is a kind of Wi-Fi fingerprint that can locate the user accurately to within a few dozen feet.

Morgan and Shean came up with a plan for the software, but because neither of them was a programmer, they hired contractors in Ukraine to actually write the code. A bigger problem was finding a way to wardrive entire cities. At first they tried recruiting Boston taxi drivers, paying them twenty dollars to merely drive their usual rounds with Skyhook's mapping gear mounted in the trunk. Everything was automated. They would install the gadget in the morning and collect it in the evening to download the accumulated data.

They immediately ran into problems. The taxi drivers would never show up at the evening meetings, or they would steal the equipment and disappear. Morgan and Shean also discovered that cabbies make lousy urban explorers, because they are creatures of habit, going to the same places every day and staying on the main roads. Left to the cab drivers of the city, vast portions of Boston would never be mapped.

They tried tow trucks; same problem. Moreover, they would have had to negotiate with hundreds of small towing companies in dozens of cities. And forget about calling the police, who patrol the same area all day. The package shipping companies, which go pretty much everywhere, might have proved a better choice. However, Morgan and Shean could not come to terms with UPS, FedEx, or the US Postal Service.

A desperate Morgan explained his plight to Shikhar Ghosh, founder of Open Market, one of the earliest Internet commerce companies. Ghosh immediately saw a solution to the problem. Hire a couple of people with cars and have them drive every street in Boston. Ghosh figured it would take three or four months at most. Morgan was skeptical, but was proven wrong almost immediately. "We paid two guys and said drive as far as you can. They completed all of downtown Boston in three weeks." Morgan was happily stunned. Creating a wireless map of every major city would be far less difficult than he had imagined, and would cost a lot less money.

Skyhook's founders easily found ambitious drivers in cities throughout the United States. However, a serious problem revealed itself one day when Morgan was cruising down Route 9, a major state highway that runs east to west across Massachusetts. At least Morgan had thought he was in Massachusetts. His location software suddenly displayed a map of Baltimore, Maryland. Either he had been teleported a few hundred miles, or something had gone badly wrong.

The answer was obvious. A Wi-Fi hot spot that had been mapped in Baltimore was now operating in Massachusetts. Its owner had moved to the Bay State and brought his Internet router and its unique MAC address with him. Morgan's mapping software knew that this particular router was in Baltimore and decided that anyone within its range must be there as well.

It was a significant problem, as many thousands of households and businesses, and their wireless routers, change location every year. Indeed, one of Morgan's earliest investors, the cell phone company

Nokia, commissioned scientists at the Jet Propulsion Laboratory to study the problem. The rocket scientists declared that Morgan's plan was doomed. So many routers would move every year that reliable Wi-Fi navigation would prove impossible.

Morgan and Shean saw a way out. Their system would require regularly "redriving" every city in order to update the maps. In the process they would pick up many signals from routers that had been in the same locations for years, but also signals from brand-new routers, or ones that had been moved. By logging the date and time when each signal was detected, the software could discriminate between "new" and "old" signals and assign greater weight to the old ones. If the software picked up five routers that had been broadcasting from Boston's Beacon Hill neighborhood for three years, it would conclude that the traveler was in Boston. A sixth router that was new in town, and had a digital address assigned to Chicago, would simply be ignored.

The software was difficult to develop, but their hired programmers came through. By early 2004, their Wi-Fi Positioning System (WPS) was consistently accurate to within twenty meters, or about sixty feet—good enough to help a traveler find points of interest like a nearby hotel or automatic teller machine (ATM). Morgan and Shean launched their new company, Quarterscope Solutions, at a cell phone–industry trade show and there received a prize for having one of the most innovative new technologies on offer.

The following year Quarterscope was renamed Skyhook Wireless, and its WPS system went on sale. Early customers included Cyber-Angel, a Nashville company that sold an antitheft service for laptop computers. With WPS on board, a stolen CyberAngel-equipped laptop could transmit its location, giving the cops a decent chance of tracking it down. Skyhook was also attracting welcome attention from investors. Begun with $1.8 million in funding from Nokia and from Boston venture fund CommonAngels, the company secured another $6.5 million in late 2005 from a group that included Bain Capital Ventures and the investment arm of chip maker Intel.

It was time to make WPS grow. That meant finding more and bigger customers. Morgan was especially eager to sign up Google. He had begun discussing his idea with search company executives in 2005. Why not use WPS to provide localized search results to travelers? A salesman spending the night in Chicago, for instance, might run a Google search for Japanese eateries. With Skyhook's WPS on board, Google would know instantly that the salesman was at the Ritz-Carlton Hotel on North Michigan Avenue. It could display names, phone numbers, and directions for the five nearest sushi restaurants. To show off the concept, Skyhook developed Loki, a program that added WPS technology to any Windows laptop. Built as a demo for Google to show that browsers should have location, Loki was later distributed free to anyone who was curious about the technology.

It worked, though not for Google. By 2007 its relationship with Skyhook had soured. According to papers filed by Skyhook in one of its lawsuits against Google, the search company asked Skyhook to hand over its Wi-Fi map database. Next to the software that made the system work, this database was Skyhook's most valuable asset, and Morgan had no intention of giving it away. "Growing skeptical of Google's motives, Skyhook declined to provide this highly confidential information to Google," the court document states. Within a few months Google unveiled its own Wi-Fi navigation service, based on a database the search company had compiled on its own, in a manner that was destined to cause Google a great deal of trouble.

In the meantime, the creation of Loki bore fruit during an early-2007 meeting between Mike Shean and Bob Borchers, a Stanford-trained engineer and an executive at Apple. Nothing seemed to come of the meeting, but six months later Morgan got a call from Borchers. Borchers asked if it would be possible to run Loki on a MacBook for a demo. He did not elaborate, but Morgan knew that Apple would soon have one of its regular meetings in which senior executives would pitch new product ideas to Steve Jobs.

Skyhook had created a version of Loki for Apple's Mac computers a few months earlier, but it was not good enough for Borchers. He had seen Loki run as an add-on program for the Firefox Internet browser, but worried that Jobs might reject a browser-based program not running on Apple's own browser, Safari. "I kind of begged him to quickly do a Safari port," said Borchers, today a general partner at venture investing firm Opus Capital, "because I knew if I showed it on Firefox, that wouldn't be popular."[8]

Meanwhile, Morgan and Shean were determined that Jobs see their technology at its best. Shean had managed to identify the resort where the executive meeting would be held. He flew to the location with a Skyhook wardriving kit, rented a car, and drove throughout the area to make sure it was thoroughly mapped.

The demo was a huge success. A few days later, Morgan checked the voice mail on his cell phone, listened for a few seconds, and hit the delete button. Obviously, it was a prank call. He knew one of Mike's friends was a commercial voice-over artist and surely able to imitate Steve Jobs's voice. Luckily, he had second thoughts and quickly undeleted the message. In fact, Jobs himself had phoned Morgan because he believed Skyhook could solve a nagging problem with its newest product, the iPhone.

Given the remarkable capabilities of today's iPhones, it is easy to forget that the original version, which went on sale in June 2007, suffered from some surprising limitations. At the time, millions of cell phones had built-in GPS technology, which enabled turn-by-turn driving directions. The giant carrier Verizon Wireless even included GPS in its cheapest phones, which came free of charge with a two-year service contract. Yet the original iPhone, which carried a six-hundred-dollar price tag, had no GPS.

It was a matter of technology. Apple had decided that AT&T would be the only US carrier to offer the iPhone, an arrangement that would persist until 2011, when Verizon became allowed to sell it. AT&T's phone network was built to a standard called the Global System for Mobile Communications. It was a logical choice; GSM was by far the most common cellular technology on earth, used by more than three-quarters of all cellular networks worldwide. Verizon Wireless used a less-common alternative system called CDMA2000 or CDMA.

As discussed in Chapter 5, in 1996 the FCC decided to require that cell phones transmit their location to 911 emergency services, in case a caller faced a life-threatening crisis. That meant adding a technology to let the phone quickly calculate its precise location. Qualcomm responded by adding GPS capability to its CDMA phone chips. It was like getting a cell phone with an advanced navigation system thrown in for free. By the mid-2000s GPS was standard equipment on millions of cheap phones from Verizon Wireless and Sprint, another major cell company that used the CDMA standard.

Makers of GSM chips did not rush to add GPS features to their silicon. Instead, GSM phone makers would have to add a separate chip, significantly raising the cost of the phone. For pinpointing 911 calls, AT&T chose a different location technology that calculated the phone's position by the strength of signals from nearby cell towers. Although not as accurate as GPS, the method worked reasonably well in emergencies.

Apple could have added GPS to the original iPhone, as it has in every version since. Yet according to Borchers, there were good reasons to hesitate. For one thing, the weak signals broadcast from GPS satellites are lousy at penetrating walls. Apple engineers reasoned that GPS would be of limited value indoors, where iPhone users spent most of their time.

Yet Apple was mainly worried about extending the life of the iPhone's rechargeable battery. GPS consumes an immense amount of power. Adding the necessary chip would significantly shorten battery life, especially if users left the GPS service running at all times. The battery issue was particularly sensitive; whereas other smartphones, like BlackBerries, made it easy to swap batteries, the iPhone's battery was built into the device and could be replaced only by a trained technician. A dead battery is a show-stopping problem for iPhone owners.

Thus, the original iPhone shipped with no GPS. Critics who otherwise hailed the phone sniped at Apple for leaving out satellite navigation, which by 2007 was becoming a commonplace tool. Apple knew it needed to bring a decent navigation technology to the iPhone, and Skyhook was ready with a software solution that could be retrofitted to the millions of iPhones already on the market. Within days Morgan met with Jobs. A few months later they signed a deal. Skyhook was on the iPhone by January 2008.

In the afterglow of the Apple deal, Skyhook continued to sign prestigious new clients. GPS chip maker Broadcom agreed to integrate Skyhook technology into its products, and Qualcomm licensed it for use in its GPS-enabled phone chips. This allowed them to offer a product that would provide location data even indoors or other places where the weak GPS satellite signal could not reach.

Morgan also hoped to profit from an indirect relationship with Google. In 2005 the search company had surprised the telecom industry by purchasing a small California software developer called Android, which was working on software for mobile phones. After two years of stealthy development, Google in late 2007 revealed that it was developing Android into a full-fledged operating system for smartphones. Android would boast capabilities to rival those of Apple's recently released iPhone. Moreover, the software would be licensed free of charge to any cell phone manufacturer who wanted it, allowing them to quickly flood the market with sophisticated yet inexpensive phones.

Although the first Android-based phone, the T-Mobile G1, was a dud, Motorola's Droid, released in December 2009, was sleek, light, powerful, and loaded with an improved version of Google's Android software. Motorola, which had not had a hit cell phone in five years, was desperate for success. The Droid's exclusive vendor, Verizon Wireless, was sick of having its customers defect to AT&T so they could use the iPhone. Between them, Verizon and Motorola spent $100 million on a massive holiday advertising campaign, featuring TV ads in the style of big-budget science-fiction films. The ads' message was bold—we've got a phone that's smarter and better than the iPhone. The claim might have been debatable, but the results were beyond dispute. According to Flurry, a telecom-industry analyst firm, Verizon Wireless sold a quarter-million Droids in the phone's first week and just over a million in its first seventy-five days. Even the original iPhone had not sold that fast.[9]

The Droid was the first Android phone to make a dent in the market, but there were plenty more in the pipeline. Morgan wanted to get Skyhook's technology on as many of them as possible. Skyhook had developed next-generation software called XPS, which embraced a hybrid approach to location finding. The software could lock onto Wi-Fi signals, but because GPS provides superior accuracy, XPS would lock onto satellite signals whenever they were available. If neither satellite nor GPS radios were in range, XPS defaulted to last-ditch mode, estimating its location by triangulating signals from nearby cell towers.

Determined not to be unduly dependent on its relationship with Apple, Morgan wanted to put the technology on as many Android devices as possible. Even though discussions with Android's parent, Google, had gone nowhere, Skyhook still held out hope, with good reason. Google had made Android an "open" system, allowing phone makers to make substantial modifications to the software. Google did establish a few basic standards with which all device makers must comply, in order to have their phones certified Android compatible. However, Morgan believed that Google had left plenty of room for Android phone makers to add Skyhook's technology.

At the same time, Google was developing a navigation service of its own. Like Skyhook, the company had decided on a hybrid approach, combining GPS and Wi-Fi location. Having failed to strike a bargain with Skyhook, Google started creating its own Wi-Fi maps. There was no handwringing about the ruinous cost of driving down every street in every city, either—with 2008 revenues of nearly $22 billion, Google could afford it.

Google was already spending millions to map the planet. In 2004 the company acquired Australian Internet map company Where 2 Technologies and American satellite photography firm Keyhole. By February 2005 Where 2's upgraded, rebranded service was launched as Google Maps, while Keyhole's detailed photographs were used to create Google Earth, a digital globe that lets the user zoom in from space to view overhead imagery of nearly anyplace on the planet.

Google spent large sums to keep these geographic resources up to date. For Google Maps, that meant maintaining a fleet of GPS-equipped cars festooned with cameras and laser range finders. As they drove streets throughout the United States and the world, these cars generated rich 3-D imagery of the places they mapped. Google began using the images to create a new kind of digital map that gave users a street-level view of a place. A user could take a virtual stroll down a street, seeing exactly what he would see if he had gone in person.

From its launch in May 2007, the new Street View service was popular with deskbound explorers, but attacked by privacy advocates. The Street View mapping vehicles captured images of individuals, some of whom did not care to have their activities put on global display. In the United States, Europe, and Asia, wherever Street View went, lawsuits and government privacy investigations invariably followed. Google made some concessions—deleting sensitive images and developing a method to turn human faces into unrecognizable blurs of color. These were small nuisances on the way to the company's goal—an eye-level, street-by-street map of the world.

Because Google's Street View cars were already on the road, it was easy to give them something else to do. The company installed Wi-Fi radios in its vehicles to collect the same kind of data Skyhook had gathered. By late-October 2008 Google had added Wi-Fi-based location to the smartphone version of Google Maps.

One year later Google announced that future Android phones would deliver turn-by-turn street navigation for free. It was frightful news for giant navigation companies like Navteq and TomTom, who make much of their revenue from subscription-based navigation services. But Google, at heart an advertising company, needed users' attention, not their cash. The company would happily give away valuable navigation data, as long as users had to pass through Google to get it.

Like Skyhook's service, Google's navigation system was a hybrid of GPS, Wi-Fi, and cell-tower location mapping. It is an open question whether Google's effort violated the patents Skyhook had earned—a question making its way through federal courts. Yet even before Skyhook filed its patent lawsuit, Google's Wi-Fi mapping campaign had spawned the biggest scandal in the company's history.

It began in Germany, where memories of Nazi and Communist rule had spawned a strong tradition of personal privacy. Google's Street View cars, covered with digital cameras, got a chilly reception early on from German citizens and government officials. In response, Google offered Germans the option to request that images of their homes be made unrecognizable. Of 8.4 million German households, about 250,000 signed up.

Still, politicians like German foreign minister Guido Westerwelle remained hostile to Street View. Germany's privacy authorities also cast a suspicious eye on the service. Apart from the company's federal data-protection authority, each of Germany's sixteen state governments has a similar agency of its own. The agency in Hamburg contacted Google in April 2010, demanding that the company provide a full accounting of all the information gathered by its Street View cars. Google complied with a minimum of fuss, but the response stunned and outraged German authorities. The company had never before told them that the Street View cars that photographed German homes were also intercepting their Wi-Fi traffic. Germany's federal privacy commissioner, Peter Schaar, demanded that Google delete all the data it had collected.

Peter Fleischer, Google's global privacy counsel, wrote in a blog posting that the company had not previously revealed the practice to German authorities because it was unrelated to the fuss over Street View's public photographs and because other companies, like Skyhook, were doing the same thing with hardly a peep of protest. Still, Fleischer admitted, "it's clear with hindsight that greater transparency would have been better."[10]

Fleischer seems to have meant what he said, because within days Google went public with a startling confession. The company's Wi-Fi mapping program was far more intrusive than it had previously admitted. Google had not simply mapped the identification codes and locations of millions of routers; it had intercepted and recorded the data being transmitted by users. Their e-mails, their bank records, their visits to Internet sites—great masses of these digital files had been captured and stored in Google databases. "Quite simply, it was a mistake," admitted Google senior vice president Alan Eustace. In 2006 a chunk of code that captured Wi-Fi data was included in an experimental program meant for use only in the lab. When Google built its software for real-world Wi-Fi mapping, the experimental code was included. Accidents will happen. Yet unintentionally scooping up six hundred gigabytes of other people's Wi-Fi over a period of four years gives new meaning to the word *mistake*.[11]

Of course, Skyhook had also scooped up digital IDs from millions of Wi-Fi routers. But Skyhook had used a call-and-response method reminiscent of a lively Baptist church service, where the pastor shouts "Hallelujah!" and the congregation chants "Amen!" A Skyhook detector broadcasts a digital call to nearby Wi-Fi routers. If any are within range, they blurt a reply that contains the router's ID code. The virtue of this method is that the simple "amen" of the router contains none of its owner's personal communications. Skyhook receives the router's unique code and nothing else.

By contrast, Google had its Street View vehicles driving the streets of cities around the world, scooping up whatever Wi-Fi transmissions came to hand. Of course, they captured the necessary digital IDs for their Wi-Fi map, but also countless fragments of personal data. Much of it would have been meaningless; sensible Wi-Fi users activate their routers' encryption software to garble their transmissions, and Gstumbler, the Wi-Fi detection software used by Google, automatically discarded encrypted data. However, Gstumbler was programmed to preserve any unencrypted data that came its way. Again, because this information was being collected by a moving car, much of it would have been incomplete fragments of little value. Still, there were plenty of chances for Google to capture credit card account numbers, bank statements, and other sensitive information. Indeed, an investigation by privacy authorities in Canada found that the Street View cars had copied full names, telephone numbers, and street addresses of Canadian citizens, as well as complete e-mail messages.

Whatever Google's motives, the Street View Wi-Fi revelation morphed into a global fiasco. In more than a dozen countries, government agencies launched investigations to determine whether privacy laws had been breached. The US FCC opened a probe; so did law enforcement agencies throughout the European Union and Asia. The governments of South Korea, Australia, and New Zealand each found

Google in violation of privacy laws; in March 2011 France punished Google with a fine of 100,000 euros. In March 2013 Google paid a total of $7 million to thirty-eight US states and the District of Columbia to settle their lawsuits over the Street View matter. For a company whose unofficial motto is "Don't be evil," the whole event was a bitter public humiliation. No surprise that in May 2010, Google said its Street View cars would no longer capture Wi-Fi mapping data.

By then Google had recruited a global army of Wi-Fi mapmakers—the millions who carried cell phones that ran the company's Android operating system. All Android phones contain both Wi-Fi and GPS chips and are constantly in communication with cellular networks. All Google needed was software that would make each phone detect incoming Wi-Fi signals and match them up to the phone's GPS location. It would work like the Street View cars, only without capturing personal data. Only the Wi-Fi router's digital ID would be intercepted.

Google avoided any suggestion that it was violating the privacy of Android phone owners. When setting up their new phones, users are asked if they are willing to reveal their location data to Google; they can refuse to provide the data feed or switch it off later if they change their minds. All incoming data are anonymized, so that even Google does not know whose phone is providing it. Even the phone's digital ID number is scrambled to make it unreadable. All Google knows is that an Android phone at a given latitude and longitude is receiving the Wi-Fi ID codes of three nearby routers. Compile enough such data points, and you can map any city on earth. That is, only if the Android phones are using Google's Wi-Fi location system. If a phone maker opted out and began using an alternative Wi-Fi mapping technology, like Skyhook, Google would lose access to that steady flow of updated Wi-Fi data. The accuracy of its maps would decay, eventually to the point of uselessness.

In Morgan's opinion, this worrisome thought occurred to top Google executives in late-April 2010—when the online news service *Business Insider* announced that Motorola would replace Google's Wi-Fi mapping system for Android phones with Skyhook's technology. The news dismayed Google management, which reacted with an aggressive counterattack that could threaten the survival of Skyhook.

Skyhook fired back with lawsuits in Massachusetts and federal courts.[12] According to Skyhook's Massachusetts court filing, the deal with Motorola had actually been struck months earlier. Skyhook had spent nearly four years and $1.5 million trying to strike a deal to put its Wi-Fi location services into Motorola phones. The companies came to terms in September 2009 and then spent more months designing and testing Android phones with Skyhook navigation.

As mentioned above, even though Android is an open-source operating system, phones or tablet computers that do not meet Google's official compatibility standards cannot wear official Android certification and are not allowed access to Google's online software store. It is the kiss of death for any product that does not measure up. Skyhook claims that Motorola ran the standard Android certification tests, checked the legal language of Google's official compatibility rules, and decided that Skyhook fit the bill. Of course, Skyhook charged a licensing fee for its product, whereas Google's was free. Yet Motorola considered Skyhook the better bargain, as they found the system delivered more accurate navigational results than Google's. Motorola's plan was to differentiate itself from other Android phone makers. The company's devices would be fully compatible with Google's standards, but would incorporate extra features that promised better performance than plain-vanilla Android.

For all the dismay expressed by Google executives, news of the Skyhook-Motorola deal may not have reached Google cofounder Larry Page until late-May 2010, a month after it was announced. According to Google e-mails submitted to the Massachusetts court, that was when Page wrote senior vice presidents Jonathan Rosenberg and Alan Eustace, demanding to know what was being done about Skyhook. In response, Lee wrote that "we absolutely do care about this because we need wifi data collection in order to maintain and improve our wifi location service (especially after having Street View wifi data collection discontinued)." In other words, Android phone data had become vital to the survival of its Wi-Fi mapping system, and Skyhook's inroads represented a serious threat. Of course, Google could simply license the technology from Skyhook. They would have had to pay for it—an unwelcome idea, as Google gives Android away at no charge. Lee wrote that Google had no desire to rely on an outside company for such a vital service. "There would be incredible risk to depend on them," he wrote of Skyhook.

Google's location tech team sprang to full alert, firing off messages to Motorola and arranging face-to-face meetings aimed at convincing the company that Google's Wi-Fi maps were as good as Skyhook's, if not better. Yet the Googlers did not stop there. They informed Motorola co-chief executive Sanjay Jha that using Skyhook as the primary location service on their phones would violate their license to use Android. Google demanded that Motorola refrain from shipping the affected phones to retail stores or cellular carriers until the issue was resolved.

Google insisted it was working in the interests of consumers and software designers. On Skyhook-equipped Android phones, any location-based software apps would use data provided by the Skyhook system. Thus, a third-party app that gives driving directions would rely on the location data provided by Skyhook. Yet much of the data from the Skyhook system would use Wi-Fi to determine location, rather than GPS, which is significantly more accurate. Google argued that the Skyhook software would thus degrade the accuracy of any app that relied on it.

It was Motorola's turn to be dismayed. The company had contracts to deliver the phones to retailers and wireless carriers. In the ferociously competitive cell phone business, any delay would mean lost sales worth millions. Besides, it wasn't fair. A Motorola engineer in the United Kingdom had just purchased a new phone from an archrival, South Korea's Samsung Corporation. Samsung, like Motorola, had quietly struck a deal to replace Android's location system with Skyhook's. But unlike Motorola, Samsung had begun delivering phones.

Once again, Google had been caught napping. Skyhook had signed a deal with Samsung in April 2010 but announced it three months later, in early July. By late June Samsung was selling the phones in Europe. An angry Motorola executive demanded to know why it was being singled out. Stephen McDonnell, the Motorola executive who managed the company's relationship with Google, asked for a waiver to ship its Skyhook-based phones, to put the company on an even footing with Samsung.

Google flatly refused. Instead, the company contacted Samsung and demanded that it halt further shipments of the Skyhook phones. Samsung and Motorola had two choices. They could ignore Google's demands and keep shipping phones. However, the phones could no longer be sold as truly Android compatible, making them far less attractive to consumers. Or they could tear Skyhook's software out by the roots and use Google's location service instead.

Samsung and Motorola chose the latter. It was the cheap, simple, quick solution. For Ted Morgan and Skyhook, however, it was a major setback. Skyhook responded with two lawsuits in one day. In state court Morgan charged that Google had used its massive market power to illegally disrupt Skyhook's contracts with Samsung and Motorola. The federal lawsuit claims that Google's Wi-Fi mapping effort violates four patents issued to Skyhook. As of March 2013, all the patent cases had been consolidated. They are expected to come to trial sometime in 2014.

Before these cases are resolved, your present smartphone will probably have been rendered obsolete three times over. It is not a problem for Google, which has plenty of money to see it through. Tiny Skyhook cannot afford a waiting game. The company must keep hunting partners looking to add location awareness to their digital devices. Most of them won't be phones. "I have no doubt that every single phone, every single laptop, tablet, digital camera, e-book reader in the next three or four years will have Wi-Fi location," said Morgan. Skyhook—or Google—can turn any of these devices into personal navigators.

Perhaps too personal. These same gadgets could be used to track every move we make, whether we like it or not.

Long Shots

At seventy thousand feet you can see a sky turned blue-black and star-studded even at high noon and, below it, the curve of the earth, stretching for hundreds of miles in all directions. And if you are carrying the right kind of camera, you can photograph the homes and factories and highways below, even the military bases, with their guided missiles tipped with nuclear bombs.

Francis Gary Powers had the right kind of camera, but on May 1, 1960, he was taking pictures in the worst possible place, high over the Russian city of Sverdlovsk. It was Powers's twenty-eighth flight in a Lockheed U-2 spy plane, and his last.

Powers saw an orange flash and felt a brutal thud of concussive air, and then his airplane began to come apart. The canopy tore free, and Powers was dragged into the open sky, held to the disintegrating U-2 only by his oxygen hose. He managed to rip himself loose and open his parachute. Powers abandoned any thought of using the poison-tipped suicide needle he had been issued, to prevent being taken alive, and furthermore he had not been able to reach the switch that would have shredded the aircraft's top-secret film and camera equipment.

The government of the Soviet Union captured Powers and his illicit photographs intact, red-handed proof that the United States was violating international law by flying spy planes over Russia. It was one of the greatest American humiliations of the Cold War.[1] Worse, it was literally a poke in the eye. The United States dared not use the U-2 over Russia anymore, so the nation's military planners were blinded, their most reliable source of intelligence lost.

But not for long. Less than five months later, on August 19, Powers was sentenced in a Moscow courtroom to ten years' imprisonment for spying. On the same day US Air Force lieutenant colonel Harold Mitchell, flying a modified C-119 cargo plane about three hundred miles west of Hawaii, swooped in on a metal and ceramic capsule dangling from a parachute. After two unsuccessful tries, Mitchell snagged the capsule at an altitude of eighty-five hundred feet. A spacecraft had been brought back to earth intact—an essential prerequisite to human spaceflight.

Ever since Russia's 1957 Sputnik triumph, the US space program had been stuck in second place. Here at last was something to brag about. The scientists, engineers, and pilots involved in the project were celebrated as heroes of the space age. They were feted with medals and parades and made guest appearances on the big TV shows.[2]

They deserved it all, though few Americans knew why. Discoverer 14's reentry capsule had been filled with photos covering 1.4 million square miles of the Soviet Union, more than all the images shot during the twenty-four successful U-2 missions over Russia. Among the images it captured were pictures of the same missile base that Powers had been sent to photograph. Discoverer was the name created for public consumption. The program's true name was Corona, a campaign, funded by the Central Intelligence Agency (CIA), that built the world's first photo-reconnaissance satellites.

A few journalists and intelligence experts guessed the real purpose of the Discoverer project; the Soviets knew as well and bitterly denounced it. Yet they could do nothing. They had no way to shoot down a satellite and no legal basis for doing so. The Soviets' own

Sputnik, the first space satellite, had flown over the United States and other countries without permission, thereby setting a precedent in international law that still stands—national boundaries end where outer space begins.

In the long run the Russians—among others—would benefit from the Corona program and the far more capable reconnaissance satellites that would follow it. Satellite photography, combined with images shot from aircraft, would make possible new maps of the earth, maps more accurate and more useful than any previously created.

People understood the potential of mapping from above long before humans had access to airplanes or spacecraft. The hot-air balloons built by the Montgolfier brothers of France in the 1780s were the first man-made flying machines. By 1794 the armies of revolutionary France were sending observers aloft to view enemy troop deployments. The practice continued in the Napoleonic era; spy balloons were used in the French invasion of Egypt in 1798.[3] Yet the value of these intelligence reports depended on the eyesight and judgment of the observers, who would rise hundreds of feet in balloons tethered to the earth and then fling written reports to the ground below.

The invention of photography in the 1820s enabled the creation of accurate, objective images of the world. Still, early photo techniques were too complex and primitive for use in the confines of a balloon. It was not until 1858 that French photographer Gaspard-Félix Tournachon captured the first airborne image, a picture of the village of Petit Bicêtre, outside Paris.

In the 1850s French Army officer Aimé Laussedat developed tools for the new science of "photogrammetry," the process of translating photographs of the world into accurate maps. By shooting photos of various landmarks from multiple locations, he developed methods for calculating the exact positions of his targets. A decade later Laussedat was testing his gear not in balloons, but on the rooftops of buildings in Paris.[4] Through aerial photography and photogrammetry, it was possible to accurately map great swaths of the world without sending in teams of surveyors.

Such maps would require the collection of millions of aerial images, far more than could be gathered by tethered balloons. And airborne photography was still too difficult to be practiced on a large scale. During the American Civil War, a balloon corps established by the Union army to spy on Confederate troop deployments gave up on cameras. Instead, the aerial observers made pencil-and-paper sketches of what they saw beneath them. The results were so unsatisfactory that the Union had given up on recon balloons by 1863.[5]

Aerial reconnaissance came into its own when the Wright brothers got airborne in 1903. Powered airplanes, capable of flying straight, level, and low, were ideal platforms for shooting pictures. Meanwhile, camera technology had advanced. The glass or metal plates once used to capture photographic images had been replaced by lightweight, flexible photographic film. New cameras had been custom built for aerial use, with multiple lenses capable of creating three-dimensional pictures of the earth. By revealing the depth and height of various landmarks, such cameras greatly simplified the process of mapping.[6]

Unfortunately, the know-how and technology for aerial mapping arrived at the worst possible moment. Many thought the European war that broke out in August 1914 would be a replay of the Franco-Prussian War of 1870—ugly and brutal but destined to end in a few months at most. Instead, it developed into the bloodiest human conflict up to that time, a four-year campaign of mass slaughter. The Great War was especially bloody due to the application of new technologies like machine guns, poison gas, and, above all, highly accurate breech-loaded artillery pieces, which could fire several shells per minute each. These big, fast guns killed more people than any other weapon of the war. One reason for the appalling accuracy of the guns was the use of aerial spotters to help aim the shells where they would do the most harm.

Well before the war began, European military experts realized that aircraft would forever alter the nature of battle. "Airplanes are also as indispensable to armies as the cannons and the rifles," declared French general Pierre Auguste Roques, after witnessing their performance in his country's military maneuvers in 1910. Similar experiments in Britain and Germany also made an impression.

However, not everyone was impressed. General Douglas Haig, destined for supreme command of the British army during the war, was defeated during his country's war games of 1912 by General Sir James Grierson, who used aircraft to track every movement of Haig's force. Despite this setback, Haig insisted that aerial reconnaissance would not amount to much in a real-world fight. In 1913 he told students at the army's staff college, "I hope none of you gentlemen so foolish as to think that aeroplanes will be usefully employed for reconnaissance from the air. There is only one way for a commander to get information by reconnaissance and that is by the use of cavalry." Haig's French counterpart, Field Marshal Ferdinand Foch, held a similar view. Both men quickly came to their senses once the war commenced and became avid supporters of air reconnaissance.[7]

The airborne observers sharply increased the lethality of artillery bombardments, as aerial photos could show gunners exactly where to aim their shells. Armies did their best to shoot down such planes. By 1915 the Germans had built the first purpose-built fighter aircraft, the Fokker Eindecker, to go gunning for the camera planes.[8] The Allies did the same, and the skies became a killing zone. By 1917 the typical recon pilot could expect to last no more than three weeks before being shot down. Britain alone was losing two hundred pilots a month.[9]

Still, the air war continued because the information obtained by the flying spies was invaluable. By the time the war came to an end, a quarter of all combatant aircraft were recon planes equipped with cameras. They shot millions of photos—gun emplacements, supply depots, and concentrations of enemy troops. Observers learned how to develop film while still aloft, enabling them to prepare their photos while the pilot was heading back to base. When the plane touched down, the observer could hand finished photos to couriers who could take them directly to intelligence officers for analysis. In 1918 American recon flyers were delivering photos of enemy targets to commanders just twenty minutes after clicking the shutters.

After the war aerial photos were still in demand by mapmakers around the world. The US Army Corps of Engineers and the US Geological Survey were major customers for the photos produced during the 1920s and 1930s by a US Army Air Service that had been much diminished by postwar budget cuts. Pioneering air photographers like George Goddard developed new methods for nighttime photography and the use of film that captured infrared radiation instead of visible light. Infrared film can often deliver clear, sharp pictures even on days when the ground is obscured by dust or fog.[10]

The Great Depression that ravaged the American economy of the 1930s proved an unexpected boon for aerial photography. The New Deal economic programs launched by the Roosevelt administration included a host of efforts to upgrade the nation's infrastructure. The Tennessee Valley Authority, for instance, brought reliable electric power to millions of people in the central United States. However, building the TVA's vast network of hydroelectric dams and artificial lakes would require a far better understanding of the area's geography. As of 1937 fully 46 percent of the continental United States still had not been mapped in detail. Aerial photographs, analyzed with the new tools of photogrammetry, could generate the needed maps quickly and at relatively low cost. A host of New Deal–subsidized programs hired aerial surveyors to shoot the needed images. By June 1938 about 1.6 million square miles of the continental United States had been photographed, or about half of the country.[11]

Still, when America entered World War II in 1941, US photo reconnaissance was little better prepared for combat than it had been in 1914. Budget cuts had shredded the military's air surveillance operations. The US Army Air Corps set up a photography school in 1938, but provided training only in how to shoot pictures from the sky, not how to interpret the images. The air corps did not even have separate aerial reconnaissance units; instead, the recon function fell under control of the bomber command.[12]

As it had in World War I, the United States gradually built up an air-recon capability. Once again, the Americans benefited from the experience of the British, who had been at war with Germany two years before the United States joined in. The Allies developed highly efficient systems for photographing millions of square miles of terrain and rapidly analyzing the images for information of military value. It was an intelligence operation of extraordinary scale. The Allied photo reconnaissance center for Mediterranean operations became the largest photo-processing center on earth, using 20,000 gallons of water and 600 gallons of processing chemicals each day and 31 tons of photo printing paper each month.[13]

Air reconnaissance played a vital role in every major mission in the Second World War. For example, camera-equipped planes accompanied the British and American bombers that rained 2.7 million tons of bombs upon occupied Europe. The hundreds of thousands of photos from these raids helped war planners evaluate the effectiveness of each strike and were useful in identifying additional targets.

These wartime photos remain valuable; indeed, they are lifesavers. A German company, Luftbilddatenbank, or "Aerial Photo Databank," retains an archive of them for use in locating unexploded bombs left over from the war. It is estimated that half the bombs dropped did not explode and that many of these are still potentially lethal. In Berlin alone there may be fifteen thousand unexploded bombs still unaccounted for. Luftbilddatenbank has digitized a vast collection of wartime photos to pinpoint likely sites for unexploded ordnance.[14]

World War II air photos are also prized by historians and archaeologists. These images cover vast tracts of land that had never before been photographed from the air. The pictures often reveal centuries-old farms, towns, and fortifications whose existence had long been forgotten and are invisible except from above. In some cases the photos document the subsequent destruction of these sites through bombing or shelling. The aerial photos are the only evidence that these places ever existed.[15]

By war's end, much of the planet had been photographed in unprecedented detail. Captured German images had recorded nearly the whole of Europe, including Russia all the way to the Ural Mountains;

US photographers in Europe and the Pacific had covered 16 million square miles of the world.[16]

In the aftermath of war, American mastery of aerial reconnaissance once again began to fade. Yet America's growing estrangement with its erstwhile ally Russia led to continued surveillance flights along the borders of the Soviet Union and its satellites, beginning in 1946. These aircraft mainly listened in on Soviet radio transmissions and tested the capabilities of Soviet antiaircraft radars. Sometimes they would deliberately "tickle" the Russians by dipping inside Soviet airspace long enough to provoke a response; the airborne spies would record the resulting radio traffic as Soviet air defenses reacted to the intrusion. It was an early form of high-tech information warfare, but it was far from bloodless. Almost forty US planes were lost in such surveillance flights by the end of the Cold War, and more than two hundred Americans were killed.[17]

Skirting the borders of the Soviet empire to intercept radio traffic was one thing; flying inland to snap photos deep inside Soviet territory was quite another. The Chicago Convention of 1944 had established that a nation was entitled to exercise sovereign control over its own airspace. Any nation could justly use force in response to an unauthorized incursion by a military aircraft. As a result photo reconnaissance aircraft were little used in the early days of the Cold War. Indeed, the Central Intelligence Agency, founded in 1947, did not even have a dedicated photo intelligence section for the first three years of its existence.

The rift with Russia would soon restore aerial reconnaissance as a major national priority. Soviet dictator Joseph Stalin had created perhaps the most secretive empire the world had ever known. It was nearly impossible to learn even the most mundane facts about the state of the Soviet military or of the economic and industrial infrastructure that supported it. The sort of data published in the newspapers, books, or government documents of other nations was unavailable except through espionage. And two world wars had proved that the best information about a foe's military capabilities could often be obtained only from above.

The United States and its allies were desperate for such information. By 1948 the Soviets had imposed communist rule in much of Eastern Europe. In June Stalin imposed a blockade on occupied West Berlin, in a bid to starve out its American, French, and British occupiers and gain complete control of the city. The blockade continued until May 1949, and although it ultimately failed, the incident left no doubt about Stalin's imperial ambitions.

In response the United States began flying covert reconnaissance missions into Soviet airspace in the spring of 1949 in an effort to gauge the state of Russian military readiness. The earliest flights were mounted in the Soviet Far East, over such targets as the Kuril Islands and Vladivostok. The daring flights spawned an aggressive Russian response, with fighters rising from their bases in an effort to shoot down the intruders. Yet time and again, the Americans escaped, relying on speed and high altitude to fly into international airspace before the Russians could catch them.[18]

Perilous as these incursions were, the United States would soon feel obliged to take even greater risks. On September 3, 1949, a US Air Force weather reconnaissance plane landed at Eielson Air Force Base in Alaska after a routine flight from Japan. Such airplanes had been equipped with air filters designed to trap the radioactive particles that would be flung into the atmosphere by an atom bomb detonation. This precaution may have seemed unnecessary; the US intelligence community had estimated that the Soviet Union would not be able to build its own A-bomb until the mid-1950s. However, a test of the plane's air filter found a dusting of radiation of a kind that could not have come from any natural source. Analysis by America's atomic scientists left no doubt. The Russians had set off an atomic bomb, half a decade before the United States had expected it. Indeed, the scientists concluded that the bomb must have been tested in late August, several days before its radioactive debris had been captured.[19]

We now know that the Russians had had plenty of help in building their bomb. They had infiltrated the American A-bomb program with spies who provided the Soviet Union with vast amounts of detailed technical knowledge, shaving years off the Soviet timetable.

But what mattered most was the fact that the United States had not seen it coming. Not only had American intelligence failed to accurately gauge Russia's progress toward a bomb; they had not even detected the Soviet test until days afterward.

On December 7, 1941, Japan had taken the US Navy by surprise at Pearl Harbor; in August 1945 the United States had repaid the Japanese with two atomic bombs. Now US president Harry Truman had to contemplate a future in which the Soviets might be capable of a hellish new kind of surprise attack, combining the tactics of Pearl Harbor with the weapons of Hiroshima and Nagasaki. How many nuclear weapons did the Soviets possess? How many long-range aircraft did they have to deliver their bombs? Were they able to mount bombs on guided missiles that could strike the United States from thousands of miles away? In the absence of hard data, he could only assume the worst. Such a policy carried grim consequences of its own. It meant the United States could not tailor its defenses to cope with the most likely threats. Instead, the ignorance of military planners would force them to spend billions to fend off dangers that often existed only in their imaginations. Such a wasteful policy would drain government resources from domestic needs while saddling taxpayers with new burdens. In the long run, it could cripple the US economy.

These worries drove the Truman and Eisenhower administrations to authorize an ongoing series of covert, illegal photo-recon missions throughout the 1950s. Many were carried out with modified B-47 jet bombers, giant swept-wing aircraft that had plenty of range and enough speed to help them evade Soviet interceptors. Still, intelligence experts knew they needed better tools for the job; some had been making the case for years.

Richard Leghorn, a military reconnaissance pilot with a physics degree from MIT, had photographed the US atom bomb tests at Bikini Atoll in the Pacific in 1946. The shocking power of the weapons convinced Leghorn that America must not be left open to a surprise attack by a nuclear-armed adversary. Faced with an aggressive rival like Stalin's Russia, Leghorn believed the United States' best hope for self-defense was a campaign of aerial surveillance that would make it impossible for the Soviets to prepare such an attack in secret. In a December 1946 speech at Boston University, Leghorn admitted that such overflights of Soviet territory would be illegal. So he called for the development of a purpose-built reconnaissance aircraft, a plane designed to fly so high and so fast that it would either go undetected or escape from Russian airspace before the Soviets could react.[20]

In 1951 Leghorn was assigned to Wright-Patterson Air Force Base in Ohio and put to work on designing surveillance aircraft capable of collecting images of the Soviet Union. He embraced some daring concepts, including the use of first-generation cruise missiles such as the Snark, a forerunner of today's pilotless surveillance drones. One zany plan featured the use of balloons that would be carried over Soviet territory by the prevailing winds. This plan was actually attempted in early 1956, under the code name Project Genetrix, but in a preview of the U-2 fiasco, a number of the balloons crashed en route and were promptly put on display by the Soviets to humiliate the United States. Moreover, Genetrix produced lousy results; of 516 balloons launched, only 47 were recovered. These photographed about a million square miles of Russian forests, mountains, and deserts, but only a few images of military significance. The results from Genetrix did not come close to justifying the bad publicity, and the project was scrubbed.[21]

Leghorn's best idea was for a new kind of manned surveillance aircraft. In an age of jet fighters, he realized that such a plane would have only one defense—altitude. By flying very high, above seventy thousand feet, the aircraft would be out of reach of many warplanes, and even those capable of reaching so high would take a long time to do it, by which time the American spy plane would be long gone. Leghorn's first idea was to modify a high-flying twin-engine British bomber, the Canberra. Indeed, the United States used camera-equipped Canberras throughout the Cold War. Yet the air force insisted on changes that made the Canberra heavier and reduced its cruising altitude.

Leghorn retired from the military in 1953. His successors at Wright-Patterson adopted his philosophy of aerial surveillance and continued working on a lightweight, high-altitude solution. They found it at Lockheed Aircraft, home of renowned engineer Clarence "Kelly" Johnson and his Skunk Works, a famously radical team of designers who created some of the twentieth century's most remarkable airplanes. The Skunk Works created the iconic P-38 Lightning fighter with its dual fuselage and the F-104, the first fighter capable of flying at twice the speed of sound. Now it went to work on creating the ideal spy plane.

Johnson and his team began work in late 1954. They began with the fuselage of the F-104 fighter and then added a long, thin wing similar to those found on unpowered gliders. The resulting aircraft was extraordinarily difficult to fly, but could cruise for twelve hours at a time at well above seventy thousand feet. It was hoped that at this altitude, Soviet radars might not even detect the slender, fragile aircraft.

The air force rejected the odd-looking aircraft, but presidential adviser Edwin Land, the inventor of the Polaroid instant photography process, recognized its potential. Land persuaded President Eisenhower to back the project. For his part, Eisenhower ordered the air force to provide technical support for the aircraft, but placed their operation under the control of the Central Intelligence Agency. "I want this whole thing to be a civilian operation," Eisenhower said. "If uniformed personnel of the armed services of the United States fly over Russia, it is an act of war—legally—and I don't want any part of it."[22] To avoid advertising the purpose of the new aircraft, it was assigned the most neutral-sounding military designation anyone could think of—U-2, with the U standing for "utility."

On July 4, 1956, CIA pilot Hervey Stockman took off in a U-2 on the plane's first incursion into Soviet airspace. He soared over East Berlin, into Poland, and on to Russia, on a course that took the plane over the cities of Minsk and Leningrad—today's St. Petersburg. The aircraft performed flawlessly, but not secretly. To his dismay Stockman saw Soviet fighters rising from air bases below, desperately trying to get high enough to shoot him down. They all failed. However, the spectacle settled any questions about the stealthiness of the U-2.

The Russians could not hit the plane, but they could most definitely track it. The United States made four more U-2 flights over the Soviet Union over a six-day period, halting the campaign only after the Soviets issued a formal protest. Still, the photos captured in that brief period seemed to justify the political risk.

Two years earlier the Russians had shown off their newest long-range bomber, the Bison, an airplane that appeared capable of dropping an atom bomb on American cities. Since then the CIA and air force had scrambled to figure out how many of these planes the Russians had. By 1956 the air force had estimated that the Russians had about a hundred of the planes, quite enough to carry out a devastating attack on the United States.

The photos from the U-2 overflights told another story. The U-2s had photographed nine major Soviet air bases. They found no evidence of camouflage; everything was out in the open, even the loading pits for hoisting nuclear weapons into aircraft bomb bays. As for the bombers—not a single Bison could be seen.

CIA analysts realized that the Bison was a paper tiger. The Soviets had built a prototype or two, but had not put the plane into mass production.[23] Instantly, the threat of a sneak attack faded dramatically. The United States could disregard Soviet saber rattling about its bomber fleet, and vast sums that might have been wasted on defensive measures could be put to better use.

Years later Eisenhower wrote in his memoirs, "Perhaps as important as the positive information—what the Soviets did have—was the negative information it produced—what the Soviets did not have. U-2 information deprived Khrushchev of the most powerful weapon of the Communist conspiracy—international blackmail—usable only as long as the Soviets could exploit the ignorance and resulting fears of the Free World." Presidential science adviser James Killian said that data collected by the U-2 saved the United States billions of dollars in needless defense expenditures.[24]

Despite the immense value of the U-2 images, Eisenhower fretted over the risks of obtaining them. He noted that if the Soviets flew spy planes over US territory, he would order them shot down and could hardly object if the Soviets responded in kind. For the remainder of his presidency, Eisenhower authorized the incursions sparingly, and only when they were expected to gain information of the highest value. For example, a mission code-named "Touchdown," flown on July 9, 1959, produced the first photos of several sites that produced enriched uranium and plutonium for nuclear weapons.

In September of the same year, Soviet premier Nikita Khrushchev visited Eisenhower at the presidential retreat at Camp David. It was a relatively friendly visit that hinted at a possible thaw in diplomatic relations. Khrushchev never mentioned the U-2 incursions, and Eisenhower concluded that the Soviets were beginning to tolerate the flights. Maybe they had realized that giving the Americans a peek behind the Iron Curtain could reduce the risk of war. In fact, Khrushchev told his son Sergei that if he had merely complained about the U-2 flights, Eisenhower might perceive the Soviets as too weak to do anything about them. Khrushchev was determined to deal with the matter in a more decisive way.

And he did, less than a year later, when the final U-2 flight over Russia came to its inglorious end. Francis Gary Powers's fateful flight was the last American photo-recon flight over the Soviet Union. Yet without the U-2 the Soviets' military activities were virtually invisible to the United States. The danger of surprise attack was multiplied; so was the risk that the United States might respond too aggressively to nonexistent threats.

Yet as one window into the Soviet Union closed, another opened. Back in 1946 a think tank at the Douglas Aircraft Company called Project RAND had issued a study predicting that the United States could put an artificial satellite in orbit by 1951, at the immense cost of $150 million. According to the report's authors, it would be money well spent. "A satellite vehicle with appropriate instrumentation can be expected to be one of the most potent scientific tools of the Twentieth Century," the study argued. "The achievement of a satellite craft by the United States would inflame the imagination of mankind and would probably produce repercussions in the world comparable to the explosion of the atomic bomb."[25] Although the study focused on the scientific and technological challenges of creating such a satellite rather than its military applications, it briefly mentioned the use of satellites as surveillance platforms for measuring the accuracy of bombing raids and tracking weather conditions over enemy targets.

In 1951 the now-independent think tank RAND Corporation went further, in a report arguing that it should be possible to build a satellite with a television camera that could soar over the Soviet Union, broadcasting images back to the United States. The quality of the images would not be terribly impressive. At best the TV technology of the time would be able to detect only objects two hundred feet or larger in size. However, the study concluded that such relatively crude images would still provide valuable intelligence.[26] A more detailed report issued in 1954 concluded that the US Air Force could—and should—build a television recon satellite. Such a system could be ready by 1960, the report said, and would cost about $165 million. The air force agreed and by 1956 had set up a program to build and launch such a satellite.[27]

Some Eisenhower administration officials were nervous about the legal implications. Could one nation lawfully orbit a satellite—a military spy satellite—over another nation's territory? Furthermore, the administration did not regard spaceflight as a high priority. Even after the CIA warned that the Soviets would be ready to launch a satellite into orbit in 1957, the administration was unconcerned; Secretary of Defense Charles Wilson said he did not care if the Russians got to space first.

So the administration took its time. In 1955 it announced a plan to launch a scientific research satellite into orbit as part of the International Geophysical Year, a global science project involving nearly seventy countries and scheduled to run from mid-1957 to the end of 1958. The US Navy was to lead the satellite program, code-named Vanguard. The administration hoped that a successful launch would establish the legal precedent that national borders did not apply to space satellites. Meanwhile, the air force spy satellite program had been put on hold.

The Soviets' launch of Sputnik in October 1957 cured Eisenhower of his complacency. It also eased his concerns about international law. The Soviets had gone first; the way was clear. Within days of Sputnik's launch, Eisenhower and Assistant Secretary of Defense Donald Quarles were talking about how to get some spy satellites aloft.[28]

By this time the RAND Corporation had soured on its plan to broadcast the images via television. They had concluded that the best equipment that could be squeezed onto a satellite would produce images too poor to be of any use. Yet there was another way—complex and technically demanding, but possible. The satellite could use a specialized type of film, capable of handling the temperature extremes and radiation found in space. As it was run through the camera, the exposed film would be loaded into a detachable canister. When the satellite had taken the desired photos, this canister would be ejected from the satellite and dropped back to earth for recovery. The resulting images would show objects as small as forty feet in size—hardly razor sharp. However, it would be much sharper than the proposed TV system and good enough to identify major military facilities and large weapons systems, such as bombers or guided missiles.

In February 1958 the administration committed itself to building a film-based spy satellite system under the auspices of the CIA, which was already running the U-2 program. To the public the satellite program was called Discoverer and was devoted to scientific research. In the CIA it was called Corona, supposedly named in honor of a CIA official's favorite kind of cigar.[29]

By this point the Soviets had launched two of its Sputniks, while the United States had yet to orbit a single satellite. The first, Vanguard 1, went aloft on March 17, 1958, and weighed a mere three pounds, less than a bowling ball. A spy satellite would weigh far more. The Americans needed bigger rocket boosters capable of hoisting such a payload. They needed cameras and film suited to the rigors of spaceflight. They had to design a reentry capsule that could bring the exposed film safely to earth. They had to do it all without pocket calculators or personal computers, but with pencils, paper, and slide rules. And they had to do it fast.

In February 1959 Discoverer 1 was launched from Vandenberg Air Force Base in California. It became the first spacecraft to be placed in a polar orbit, circling the earth from north to south. That is the ideal orbit for a recon satellite, because as the earth rotates beneath the satellite, it passes over a different region of the planet on each orbit.

The first Discoverer was a test launch with no camera or film aboard. Over the next sixteen months, the CIA would loft twelve more of the satellites. Each failed. Sometimes the rocket failed and the spacecraft never made orbit. More often, the complex system for aiming and ejecting the reentry capsule misfired, so that the film was never recovered. It was exhausting, dispiriting work. There was, however, no turning back. Eisenhower, desperate for a way to see inside Russia, was now committed to the program. His bet paid off just in time. After the first successful Corona flight in August 1960, the United States would launch 131 more between 1960 and 1972. The photos they shot were vital to US military planning during the Cold War.

Early on they demolished the idea that the Soviets had built up a huge stockpile of nuclear-tipped missiles aimed at the United States. John F. Kennedy's warnings about this "missile gap" had helped him defeat Eisenhower's vice president, Richard Nixon, in the 1960 campaign. The satellite photos proved that the Soviets had only a handful of operational missiles. The satellites also improved the prospects for arms control agreements. American leaders could be confident that with recon satellites regularly passing overhead, the Russians would be incapable of hiding any major treaty violations.

During the Cold War the United States did not openly discuss the use of spy satellites, to avoid giving away details of how well they worked. However, the Russians certainly knew about them. Within months of the first successful Corona flight, G. P. Zhukov, Deputy Chairman of the Soviet Space Law Commission, publicly denounced spying from space. By 1963 the Russians had their own spy satellites, and leaders of both nations fell silent on the subject.[30]

Corona had its limitations. Each satellite had a short life span, limited by the amount of film it could carry. The early satellite could detect only objects forty feet or larger in size. And although image resolution got much better over time—later versions could spot objects as small as three feet—such photos still missed many vital details. Worst of all, Corona images were days old by the time policy makers saw them. For example, in 1967 a Corona satellite caught images of Soviet troops massing on the border of Czechoslovakia. By the time the film had been returned to earth, developed, analyzed, and provided to the Johnson administration, the Russians had already invaded.[31]

What the United States needed was something like the RAND Corporation's original vision—a satellite that captured images electronically and broadcast them directly to earth. It was not until 1976 that such a satellite could be created, thanks to a technical breakthrough that was destined to capture a Nobel Prize and to restructure an entire global industry.

In 1969 a pair of scientists at Bell Labs, William Boyle and George Smith, were working on electronic memory devices for computers. They came up with the "charge-coupled device," a new kind of silicon chip coated with thousands of tiny imaging elements, or pixels. Each pixel on the CCD could record an incoming photon of light, transforming it into an electrical charge. The accumulated charges could be rendered as a stream of digital data, with numbers indicating the intensity of light striking each pixel. A computer could convert this string of numbers into an image. At the time millions of people used Polaroid cameras to produce photos that could be viewed in sixty seconds and called it "instant photography." Boyle and Smith were about to give the world the real thing—devices that could take a picture and display it a split second later.

Digital photo technology would eventually drive Polaroid into bankruptcy. The same fate awaited imaging titan Eastman Kodak Company, an ironic twist considering that Kodak introduced the world's first digital camera for consumers in 1975. The CCD and its follow-on technologies have supplanted film in almost every photographic application. High-resolution instant imaging is now ubiquitous and dirt cheap; sharp digital cameras are now a standard feature in billions of pocket phones.

In the mid-1970s few people were ready for digital imaging. However, the National Reconnaissance Office (NRO), which had managed America's spy satellites since 1960, could hardly wait. With CCD technology, there would be no more delays waiting for a satellite's film bucket to fill up, no more risky capsule reentries, no waiting to develop the film and pass it up the chain of command. The United States could build satellites that would beam the images to earth within minutes of shooting them. It was not quite real-time imaging, because the data took time to travel over the radio channel. But it was pretty close.

In 1976 the first satellite based on the technology was lofted into orbit. The school bus size system, called KH-11, has been sharply upgraded over the years, but its basic design remains the same. At its heart is a large reflector telescope, similar to that used for astronomical research on NASA's Hubble Space Telescope. Captured images are focused onto a high-resolution CCD, digitized, and relayed to a network of support satellites, which in turn send the data to earth. The NRO won't provide exact details on the quality of the KH-11's digital eyesight. But outside experts estimate that the newest versions of the satellite, sometimes called the KH-12, can spot objects as small as ten centimeters, or just under four inches.[32]

Such remarkable precision does not come cheap. The NRO will not disclose how much each satellite costs, but in 2009 former senator Kit Bond of Missouri sent a letter to then US director of national intelligence Dennis Blair, complaining that just one of the satellites was more expensive than the navy's latest Nimitz-class aircraft carrier, which had cost $6.35 billion.[33] The massive price tag ensures that the United States can never operate as many spy satellites as it might wish. Large swaths of the planet are literally out of sight, due to a lack of coverage. US birds are extremely expensive because they are designed with special features—supersharp optics, of course, and also a lot of extra rocket fuel. These additions allow controllers on the ground to shift the satellites' orbits and bring different parts of the planet into view.

Yet for many applications these costly options are not necessary. Nor are the US military and intelligence communities the only potential customers for satellite images. Commercial mapmakers, land developers, oil exploration companies, and civilian government agencies, among others, have provided plenty of work to companies that shoot high-quality images from airplanes. There would surely be a ready market for similar photos shot from space.

The United States was not going to share its highest-quality satellite images or the technology used to shoot them. However, early on, NASA began exploring civilian applications of space photography. The first weather satellite, Tiros I, launched in 1960, beamed down images of cloud patterns. Tiros I used the TV technology that the CIA had rejected as too fuzzy for its spy satellites. One glance at the Tiros pictures shows why; the images are too indistinct to show objects of military interest. They were, however, good enough to identify an impending hurricane. NASA followed up with a series of Tiros and Nimbus satellites, with constantly improving image quality. The GOES weather satellites now in service can distinguish objects down to 1 kilometer in size, which is adequate for watching weather patterns, if not much else.

In 1972 the United States launched the Earth Resources Technology Satellite, or ERTS-1, designed to shoot pictures of the earth for use in environmental research, land-use management, and urban planning. The concept had been proposed in 1965, but met with opposition from the intelligence community. As with weather satellites, ERTS-1's images would be made available to scientists and the general public. The new satellite used better imaging technology, capable of producing photos with a resolution of 80 meters, or 262 feet. That is still far from military grade, but dramatically better than anything previously available to civilians.

Soldiers and spies tried to block the launch of ERTS-1. They feared its pictures would reveal far too much about US satellite technology and help the Soviets devise better ways of camouflaging their activities. Others in government insisted that the scientific and commercial benefits outstripped the danger. The US Department of the Interior forced the issue by publicly announcing that it would build such a satellite on its own, to assist in managing the country's natural resources. Although the agency had neither the budget nor the technical skill to launch satellites, the public declaration finally goaded NASA into action.[34]

In the end, ERTS-1, renamed Landsat 1, lived up to its backers' hopes. In the four decades since, the Landsat series has generated millions of images of the planet's surface for sale to scientists, businesses, and governments around the world. Even the Russians would eventually become major customers. The most recent of the series, Landsat 7, has been in orbit since 1999. It photographs one-quarter of the planet every sixteen days, capturing objects as small as 15 meters, or about 50 feet, making its pictures almost as sharp as those from the early Corona spy satellites.

From early on Landsat's backers had hoped to turn satellite imagery into a commercial enterprise. They hoped that private-sector companies would be willing to pay for the photos, thereby covering the cost of the program and sparing the nation's taxpayers. The Reagan administration even tried to privatize the Landsat system, but a host of financial and political disputes eventually scuttled the plan.[35] NASA and the US Geological Survey now jointly manage the system. And as with the GPS satellite network, the United States has abandoned any thought of making Landsat pay for itself. The federal government used to charge $600 for an image shot by Landsat 7; as of 2009 all Landsat pictures are available free of charge.

Other governments have not been so generous. During the 1980s the French government launched its own photo satellite program with assistance from the space research agencies of Belgium and Sweden. In addition, the French created SPOT Image, a commercial enterprise that would sell the satellite images to all comers; at start-up in 1986 the pictures cost $155 and up. Of course, the United States was already selling Landsat images, but SPOT was no Landsat. The French satellite used technology similar to that of the American KH-11s and produced images with a resolution of 10 meters, or 33 feet—better than pictures shot by America's old Corona satellites. These pictures, more than adequate for planning a war, could be purchased by any nation or individual with a credit card.

SPOT Image sent a chill through the US intelligence community, but the latter could do nothing to stop other countries from shooting and selling spy photos. After all, the US Landsat service had set the precedent. Moreover, SPOT Image photos proved an unexpected boon to American spymasters, as SPOT Image was happy to sell its photos to the CIA and Pentagon. America's spysats, while superb, were few in number and often in the wrong places to photograph crucial activities overseas. The SPOT satellite gave the United States an extra set of eyes.

SPOT proved its value to the United States during the Persian Gulf War of 1991. Images from the French satellite and even the relatively crude US Landsat system were used to remap remote areas of Iraq and Kuwait. The images themselves could be shared with allied troops and governments; unlike the supersharp pictures from American KH-11 satellites, the SPOT and Landsat photos were not classified information. In addition, SPOT agreed not to sell its photos to the Iraqi government, thus ensuring that Saddam Hussein had no advance warning of the massive American "left hook" attack through Saudi Arabia that crushed the Iraqi army in Kuwait.[36]

In the aftermath of the war, President George H. W. Bush signed a law aimed at developing a commercial imaging satellite industry in the United States, to preserve America's lead in the technology and to compete against overseas rivals like SPOT. The first company to win a license under the 1992 law, Worldview Imaging of Oakland, California, was founded by Walter Scott, a scientist at Lawrence Livermore Laboratories who had previously worked on defensive systems for shooting down incoming nuclear warheads.

Whereas other commercial photo satellite ventures of that era were funded by massive defense contractors like Raytheon and Lockheed Martin, Worldview secured its start-up financing from Silicon Valley venture capitalists. The plummeting cost of digital electronics would allow the construction of relatively cheap satellites, while the growth of commercial space-launch companies in France, the United States, and even the former Soviet Union would make it cheaper to get the satellite aloft.[37] In 1997 the company's first satellite was flung into orbit on the back of a converted Russian ICBM. Early Bird 1 was capable of spotting objects as small as 3 meters, or 10 feet—or it would have been if the satellite had not malfunctioned after launch.

Four years later the company launched an even more advanced satellite called QuickBird, which is still in use today. QuickBird can identify objects as small as 60 centimeters, or 24 inches, across. This is ample resolution for commercial applications and for most military uses. Since then, under its current name, DigitalGlobe, the company has launched two more satellites with even better resolution.

Scott had expected private companies to be his primary customers; instead, it has been the US government, which uses DigitalGlobe photos for everything from planning disaster relief to fighting the wars in Iraq and Afghanistan. One reason is that DigitalGlobe's three active satellites can cover so much of the planet so quickly. Because the company has built ground stations in eleven countries, its satellites can photograph almost any spot on earth and deliver the image in twenty minutes or less. In about 45 percent of the world, photos can be produced in real time.

DigitalGlobe has prospered despite intense competition from GeoEye, a Virginia-based company that launched its first satellite in 1997. In 2006 GeoEye acquired still another satellite company called Space Imaging, whose Ikonos satellite, launched in 1999, was the first commercial satellite capable of picking out objects as small as 1 meter, or 3 feet. GeoEye has since launched another satellite that can see 16-inch objects. Like DigitalGlobe, GeoEye's chief source of revenue comes from sales to the US government. In mid-2012, confronted with

military budget cuts that will sharply reduce purchases of satellite images, the two companies announced plans to merge.

Between DigitalGlobe, GeoEye, SPOT Image, and hundreds of
aerial photography companies, the entire planet has been imaged
with remarkable fidelity. These images, translated into highly accurate maps, have shown us the world at a level of detail never before
possible. But lately we have learned that cartography is too important
to be left entirely to cartographers. Today's Internet-hosted digital
maps have given rise to a new generation of amateur mapmakers with
two big advantages over the professionals—there are thousands of
them, and they are everywhere.

A Map of One's Own

UNTIL RECENTLY, THE BEST-KNOWN SATELLITE IMAGES OF NORTH Korea showed next to nothing. They were nighttime images, shot by weather satellites or commercial space cameras for hire and readily available online. In these pictures you can spot national boundaries by the blazing electric lights that outline them, lights so plentiful that they show up in photos taken hundreds of miles up. South Korea, China, Japan, and Russia are aglow with electric power. In their midst sits North Korea, a nation roughly the size of Pennsylvania, almost entirely submerged in darkness. It is the full measure of North Korea's desperate poverty conveyed in a single stark image.

Of course, there were daylight satellite photos of North Korea as well, but these drab, barren images contained little to capture the imagination of anyone other than intelligence analysts or military planners. That changed in January 2013, when the Internet search company Google unveiled a new photographic map of the famously secretive nation. Geography buffs could read the names of streets in the capital city of Pyongyang and gaze at the vast monuments erected to the nation's leaders. They could even view the concentration

camps—four of North Korea's notoriously brutal prisons where countless thousands have suffered and died.

The North Korean government had not suddenly abandoned its
paranoid isolation. It had not sent legions of workers to create new
maps of the nation and share them with the world. The new details
had been provided by citizen volunteers who had combined satellite
images with eyewitness accounts from defectors to help create new
maps of their nation. Whether its reclusive leaders liked it or not,
North Korea was being remapped, as part of an ongoing global enterprise that recruits amateur cartographers in a bid to refresh and
perfect our pictures of the planet.

Google's efforts built on the work of others who had been filling
in the map of North Korea for years. Joshua Stanton, for instance,
now an attorney in Washington, served as a US Army judge advocate
in South Korea from 1998 to 2002. To his surprise, Stanton found
that most South Koreans were indifferent or skeptical about reports
of North Korean abuses. If they heard about it, they would dismiss
it as propaganda. Stanton also felt that international human rights
organizations like Amnesty International were neglecting the issue.
He decided he had to do something. In 2004 Stanton launched One
Free Korea, a blog that tracks human rights abuses in North Korea. Since 2006 he has combined refugee reports and satellite images
to pinpoint the locations of North Korean prison camps, displaying
the images on his website. "The most comprehensive mapping of the
camps is the one you'll find on my site," Stanton said.

Another longtime North Korea watcher, Curtis Melvin, started
publishing maps of North Korea in 2007 at his North Korea Economy
Watch website. Melvin, a doctoral student in economics at George
Mason University, claims that much of the North Korean data on
Google Maps is based on work he did between 2007 and 2009. "I can
tell because they are uploading incorrect data that I published back
then," he said.

Stanton and Melvin remain unknown to all but the most avid
North Korea buffs; it took Google's involvement to alert the world to
this exercise in guerrilla geography. Still, do-it-yourself mapmaking

is no longer an esoteric hobby. Much as we do with *Wikipedia*, many people routinely submit corrections and additions for the digital maps that guide us through the world. Meanwhile, a global network of amateur cartographers is well on the way to building a complete map of the planet from scratch. In the midst of natural or man-made disasters, first responders and activists, equipped with a new arsenal of digital aids, are capturing the world's attention with up-to-date maps of the danger zones. Few of these amateur mappers have degrees in geography; most could not operate a plane table or a surveyor's transit if their lives depended on it. Yet armed with new digital tools, they can leave their mark on the map.

Indeed, many of us do it every day without even realizing it. The smartphone in your pocket knows where you are, thanks to GPS, and it constantly broadcasts your location. With the location data from a hundred such phones, each sitting in a car driving down the interstate, Apple, Google, and other software vendors generate a highly accurate map of traffic during morning rush hour. Multiply by millions of users, and you can map traffic throughout the United States, or in cities around the world. Meanwhile, each phone's Wi-Fi chip seeks nearby wireless Internet hot spots, pinpointing their location via GPS. Multiply by millions, and the result is a Wi-Fi map that can guide travelers through urban skyscraper canyons, where GPS signals can't reach.

Today, humans map and remap the planet, moment by moment, constantly refreshing and enhancing our view of the world. This great shift has been made possible by digital computing. For most of us, a map is an image printed on paper or displayed on a computer screen. However, the unequaled ability of images to convey vast amounts of information at a glance can blind us to the essence of what is being conveyed. At bottom, a map is a database—a visualized depiction of information that could be presented equally well in a computer spreadsheet as a collection of numbers.

It seems like an obvious insight. For centuries people have had a numeric concept of geography, knowing that every point on earth can be described in terms of latitude and longitude. For example, the Eiffel Tower is located in Paris, France. More specifically, it is at

48.8584 degrees north latitude and 2.2946 degrees east longitude. Likewise, there exists a coordinate for every person, place, and thing on the planet. In addition, every object on the map possesses important characteristics, which can also be described with numbers. A given neighborhood of a city might contain one hundred thousand persons, with an average age of forty-two. The residents might be 32 percent black, 38 percent Hispanic, and 30 percent white. Perhaps 53 percent of them own a car less than five years old, another 40 percent own older vehicles, and 7 percent have no car at all.

Because these characteristics belong to people located in a particular place, you can now draw a map of that place displaying data about racial and ethnic makeup, the age of the residents, and their use of automobiles. A city planner could use such a map to make better decisions about how much to invest in aid for senior citizens or whether it is more efficient to upgrade the roads or purchase new buses for the public transit system. A politician might use the same map to make educated guesses about how best to win local votes.

Images that combine geographic information with layers of data related to people, places, or things in the same area are called thematic maps. Humans have been making them for centuries, and it is easy to see why. A thematic map can teach you a great deal about a region at a glance. To make a thematic map, you must start with a reasonably accurate geographic map. Then you overlay this map with "georeferenced" data, which are facts about some aspect of the world that are associated with a specific place. For instance, knowing which way the wind will blow is vital data for those who rely on sailing ships to get around. The renowned astronomer Edmond Halley collected thousands of wind measurements taken at various places on the world's oceans and transcribed the data onto a map. The result was the first map of global wind patterns, a valuable aid to navigators published by Halley in 1686.

In the seventeenth century European scholars understood the value of collecting and cataloging data on every imaginable subject. As their databases became deeper and more geographically precise, cartographers painted layers of valuable georeferenced information onto more and more maps. German mapmaker August Crome created an economic map of Europe in the late eighteenth century showing where key industrial commodities were produced. In 1798 a New York physician named Valentine Seaman created possibly the first map to depict the incidence and spread of a disease—in this case yellow fever, at the time rampant in Lower Manhattan. In 1801 the first major map of the geology of England and Wales was published, and in 1854 British physician John Snow gained national renown for his map of a cholera outbreak in London. By mapping the homes of the victims and the nearby sources of drinking water, Snow proved that the disease had been spread by the water from a particular public well, which had been contaminated with human waste.[1]

No one questioned the value of thematic maps, only their cost. Building them required months or years of daunting effort to compile the needed data and place it accurately on the map. And combining multiple themes into a single map was even harder. People created a vast array of thematic maps for specialized purposes, but the great expense of the effort always limited the concept's full potential.

A shift occurred in the 1960s, as digital computers became a viable technology. Roger Tomlinson, a British geographer who had emigrated to Canada, was approached by government officials eager to make the first complete inventory of the country's natural resources. Their goal was a set of maps covering a million square miles, displaying farms, forests, and wildlife habitats. However, producing the maps by hand would require about five hundred trained geographers working for three years.

It occurred to Tomlinson that all the relevant information could be stored in a digital database, as could the geographic information of the map. A set of digits might represent a wheat farm; another clump of numbers could stand for a pine forest. Associate the resource codes with the correct latitude and longitude data, and you could build your map of Canada inside a computer's memory. Such a map could be whatever the user needed it to be. If someone wanted a thematic map of old-growth forests in British Columbia, the computer would simply combine the necessary data and print the map. If someone else was studying waterways in the same area, she could call up a map that ignored forests and showed only lakes and rivers.

The Canadian bureaucrats signed off on Tomlinson's plan, which became the first computerized "geographic information system." An idea so good was bound to migrate; Harvard University's Laboratory for Computer Graphics was creating its own GIS software by the mid-1960s. One graduate of the program, Jack Dangermond, would found ESRI, a major commercial vendor of GIS software.

At roughly the same time, the US Census Bureau realized that with computerized maps of streets and buildings, they would be able to hold future census counts far faster and at lower cost. The agency developed DIME, a digital representation of 276 metropolitan areas containing about 60 percent of the US population. This database was dramatically upgraded in the 1980s, when the Census Bureau and the US Geological Survey developed TIGER, a digital map of the entire United States, with precise location data for thousands of counties and cities and millions of streets and houses.[2] TIGER is in the public domain, so any cartographer can use its data as a baseline when building their own maps of the United States—and virtually every cartographer does.

Among those mapmakers was a company called GeoSystems Global of Lancaster, Pennsylvania. This spin-off from the Chicago printing firm R. R. Donnelley & Company had been using GIS technology since the 1980s to print specialized thematic maps for businesses. By the 1990s GeoSystems, like Delorme, MapInfo Corporation, and the venerable mapping firm Rand McNally, was delivering many of its maps in digital form, on optical disks. However, GeoSystems stole a march on its rivals by combining GIS with the new global computer network, the Internet, to make a radically new kind of map.

At the time a New York City resident trying to get from his apartment in Harlem to a friend's house in Queens might trace the route in pencil on a flimsy paper map or bulky road atlas or squint at a pageful of confusing handwritten instructions. GeoSystems' GIS software, by contrast, could build a street map of New York City, add georeferenced data on the traveler's starting point and final destination, calculate the shortest path between the two points, and display the results in seconds. The traveler could simply print out the results and hit the road.

GeoSystems called their Internet-based service MapQuest. Within months of its launch in 1996, it had attracted hundreds of thousands of users and become one of the early Internet's best-loved sites. GeoSystems carried ads on the site to generate revenue and sold advanced MapQuest services to its business customers. Noncommercial users, however, could get all the maps they wanted, generated in seconds and free of charge. In 2000 America Online (AOL), then one of the nation's chief Internet service providers, purchased MapQuest for $1.1 billion.

Several companies, including Microsoft, launched MapQuest imitators, but none was more successful than Google. In 2004 the company bought an Australian mapping startup called Where 2 Technologies. It was the simplest way to capture the talents of Where 2's employees and especially its founders, two brothers from Denmark, Lars and Jens Rasmussen. Backed with ample Google capital, the Rasmussens built their original product into Google Maps, a superb online mapping service that has overtaken AOL's MapQuest as the Internet's favorite geoservice.

To understand Google Maps' dominance, don't visit the website. Instead, visit a company like Zipcar, which uses a Google map to display locations where its vehicles can be found. Or browse the countless personal websites and blogs where people attach Google maps of their homes, their favorite vacation spot, the location of the church picnic.

Google was not content with handing out free geographic information on its own site. The company extended the giveaway to thousands of other sites through software called the Google Maps Application Programming Interface, or API. Freely distributed to website developers, this chunk of code grants access to Google's map servers. A developer can use the API to generate a "mashup"—a mixture of digital data from two or more sources that results in something new and useful. Maps are ideal raw material for a mashup—"Where is it?" is one of mankind's favorite questions. With the Google Maps API, a website can easily include an embedded map that points visitors to the corporate headquarters, the nearest branch office, or the church where the wedding takes place in two weeks.

Website developers were enthralled by the Google service. In 2012 the company estimated that 800,000 websites make use of the Google Maps API.[3] And although most people don't know an API from a chimpanzee, they certainly noticed the excellent maps that appeared on so many of their favorite sites, each prominently displaying the Google brand. It was advertising at its most effective. Generally, you saw a MapQuest map only when you visited the MapQuest website. Google's maps were everywhere. Little by little, Google Maps gained market share from MapQuest. As of April 2012 Google Maps was America's leading travel website, with 79 million visitors that month. MapQuest was a distant second, with 29.5 million visitors.[4]

Still, the launch of Google Maps hardly shifted the earth on its axis—MapQuest had been serving up Internet maps since before Google was born. That would come a few months later, when Google launched a second geographic product: Google Earth. The inspiration for this second product had come years earlier, in the laboratories of Silicon Graphics, Inc. Barely remembered today, SGI was once renowned for its successes in building computers to generate extremely sharp three-dimensional images. This is not the kind of 3-D you see at the movie theater from behind thick black glasses. Such 3-D images seem to leap from the screen. SGI was working on the kind of 3-D you see in a good computer game, where the images recede into the screen and draw us into a visual landscape where the objects possess depth and distance. Nowadays, our home computers and game consoles can easily display such images. However, in the 1980s and much of the '90s, nearly all computer graphics lay flat on the screen, as lifeless two-dimensional renderings.

The engineers at SGI did a lot of work to change this, and their efforts propelled the company to stardom. The shape-shifting liquid metal robot assassin of the 1991 movie *Terminator 2* and the utterly credible dinosaurs of 1993's *Jurassic Park* were created inside SGI's machines. SGI's fortunes would fade; within a few years cheap home

computers would become as powerful as the company's costly work-
stations. Yet Phil Keslin, Chikai Ohazama, and Mark Aubin, a trio
of SGI engineers, would soon find themselves working on a consumer
product that would require the same kind of graphics capability.

The engineers had been working on a way to demonstrate the remarkable ability of the latest SGI computers to generate moving 3-D
images. They ended up creating a demonstration program that showed
an image of the earth from space. Starting with this God's-eye view,
the image zoomed downward, closer and closer to the earth. The images, derived from commercial satellite and aerial photos, grew steadily
larger and closer to the ground, aiming at the Matterhorn, a mountain
in the Alps. On the ground was a Nintendo gaming console. The image zoomed in even closer, inside the machine, to the SGI-designed
graphics chip inside. And then the entire process ran in reverse, zoom-
ing out again, higher and higher and once more into space.[5]

Those who saw the demo were appropriately awed. But what could
it be used for? For Keslin, Ohazama, and Aubin, the answer lay in the
remarkable ability to render an entire moving image of the earth in
real time and to zoom in for a closer look at any part. They had the
crazy idea that people might like a software program that let them
do that. The three SGI engineers teamed up with Avi Bar-Zeev, a
veteran designer of virtual reality software for the Walt Disney Company, and set out to build a business around their idea.

Their efforts came to the attention of John Hanke, a self-taught
computer programmer who had served a stint with the US Foreign
Service in Burma. Hanke was equipped with a liberal arts degree from
the University of Texas, a master's degree in business administration
from the University of California at Berkeley, and a knack for launching new businesses. In 1996 Hanke and some fellow students began
a company that built the first Internet-based adventure game to use
sophisticated 3-D graphics—the precursor of the popular game World
of Warcraft. Hanke sold the company, launched another online game
business in 1998, and sold it in 2000.

Hanke was ready for a new venture. When he saw the SGI demo
and learned that its creators were launching a company, he jumped

at the opportunity. Hanke and his new partners launched Keyhole in 2001. Named after the code word for the US government's series of KH spy satellites, Keyhole received its initial funding in part from In-Q-Tel, a venture fund set up by the Central Intelligence Agency to finance businesses whose products could benefit national security. Still, "it was a struggle in the beginning," said Hanke, "because we as a startup weren't in a position to fly our own planes or launch our own satellites." Instead, the company scooped up as many aerial and satellite images as it could afford, from every possible source.

The result of their efforts, a program called EarthViewer, enabled customers to view moving images of the earth that scrolled across the screen as smoothly as a big-budget movie. The images were real—photographs taken from space or by airplanes, photos that revealed remarkable details about the land and buildings and sometimes the people below.

EarthViewer sold for $599 per year to businesses and government agencies; consumers could buy a stripped-down version for about $80. However, Keyhole's most influential customers were the big TV networks. When the United States went to war in Iraq, you could hardly switch on a TV set without seeing EarthViewer images of the battlefield.[6]

Among those impressed by the remarkable images were the leaders of Google. They bought Keyhole in 2004 and immediately slashed the price of the consumer version of EarthViewer to $30. The search company's ultimate goal was to drop the price of online aerial maps to zero. Google was also determined to upgrade the quality of Earth-Viewer. In its cash-starved infancy, Keyhole had invested in costly high-resolution images of the United States and some other regions of the world, but used lower-quality images for the rest of the planet. Google cofounder Sergei Brin wasn't having it. "Sergei's reaction was, I don't understand why we're limiting ourselves," Hanke recalled. "I really think we ought to try to get all of it. I think we ought to try to cover the whole world."

Ever since, Google has gone to extraordinary lengths to sharpen its images of the planet. In 2008, for instance, the company helped finance the launch of a new photo reconnaissance satellite operated by the commercial satellite company GeoEye. The satellite is capable of photographing objects as small as 1.5 feet in size, so sharply that the US military and intelligence communities are major customers. Moreover, the images have generated improved maps for Google, and only Google. Google's cash infusion bought it exclusive access to the satellite's images for commercial mapmaking purposes. Rivals like MapQuest and Microsoft must buy their photos elsewhere.[7]

In April 2005, two months after the launch of Google Maps, the company added Keyhole's satellite and aerial images to the service. Searchers no longer had to settle for old-style sketches of roads and landmarks. With a click of the "satellite" icon, Google instantly displays a bird's-eye photo of the place, made of aerial images stitched together and perfectly aligned with the underlying map.

It was not the first time that aerial images of the earth had been available online. Microsoft's TerraServer project began selling images shot by Russian satellites in 1998. Yet whereas TerraServer offered static images, Google Maps fully exploited the power of twenty-first-century desktop computers and broadband Internet. A user could zoom close to his target or pull back for a wider view. Or he could slide the image north or south, east or west. The images glided across the screen with barely a flicker, as if the computer really was cruising above the earth. At the time no other online mapping service could match it. And Google Maps provided these remarkable capabilities at no charge.

Two months later Google released an improved version of Keyhole's EarthViewer as a free Internet download and renamed it Google Earth. The software, which ran on the user's own computer instead of the Internet, was a three-dimensional model of the entire planet, made up of millions of aerial and satellite photos. Once again the user could zoom in and out and rotate the image, but this time as if he had the whole world in his hands.

Not everyone was awed by Google Earth. Barry Diller, a media mogul who had helped establish the Fox and USA television networks and led the online search service Ask.com, considered it little more than a gimmick. "After you've seen your house and all those other buildings that look like toothpicks from that height, what do you do?" he asked.[8]

Diller was practically alone in his skepticism. Since its release Google Earth has been downloaded more than a billion times. Its visual appeal is irresistible. Yes, a user can look at his own house from space—but also at the Grand Canyon, the Parthenon, or the Great Wall of China, anyplace one might dream of visiting, rendered with lifelike clarity.

In addition, Google Earth let anyone create customized maps of favorite places or subjects. Hanke's team had developed Keyhole Markup Language, a simple set of codes for highlighting locations and objects on Google Earth or Google Maps. It is no accident that KML sounds a lot like HTML, the standard language for building Web pages. They are quite similar in structure and syntax. And like HTML, KML is relatively easy to use. Just as it is possible for a small child to learn enough HTML to create a decent website, anybody can cobble together a simple Google Maps overlay in KML. You don't even have to learn the language; Google has created automated tools that make the mapping process as simple as dragging the mouse pointer and clicking a button. From these movements Google generates a KML file that the user can save for later or distribute to friends and colleagues. A KML user can even create an automated visual tour that scrolls across the screen like a movie, complete with musical soundtrack.

With KML a student could create a Google Earth overlay showing where to party in Fort Lauderdale during spring break, a veteran of the wars in Vietnam or Iraq could mark the places he had fought, a historian of the civil rights movement could create a guided tour of the greatest landmarks of the struggle. With their simple tools for adding new landmarks, Google Earth and Google Maps had set the stage for a new kind of cartography—quick and dirty mapmaking that would help resolve conflicts and save lives.

In the wake of disaster—whether natural or man-made—the maps of a nation must be redrawn, and quickly. In a military crisis, like the 2011 overthrow of Libyan dictator Mu'ammar Gadhafi, foreign governments and aid agencies need up-to-date maps to track the movement of refugees. After a hurricane or earthquake has toppled buildings and swept away roads, existing maps of the ravaged area are often useless. Arriving aid workers, many of them strangers to the region, are in desperate need of accurate maps.

The act of creating such maps is called "crisis mapping," and until recently it was the exclusive domain of governments and well-funded international organizations. For example, since 2003 the United Nations has operated the UNOSAT Humanitarian Rapid Mapping Service, an agency that deploys teams of expert mapmakers to cover areas of the world ravaged by natural disasters and military conflicts. The agency uses a variety of costly resources in its remapping efforts, including commercial satellite images and photos taken from unmanned aerial drones. As of 2012 the UN service had generated crisis maps used in more than two hundred emergencies, including the Libya crisis and the 2011 Japanese earthquake and tsunami.[9]

Another leading source of crisis maps, the British relief organization MapAction, uses a mini data center, including high-end mapmaking software, computers specially modified for light weight and high performance, and printers to crank out hundreds of paper maps for first responders.[10] MapAction's first disaster mission came in December 2004, when team members deployed to Asia in the aftermath of the Indian Ocean earthquake and tsunami that killed 230,000 people. Since then the agency has responded to dozens of crises, ranging from an earthquake in China to a typhoon in the Philippines.

Vital as these services are, they are limited in their reach by their dependence on experts. A team of skilled mapmakers can accomplish a lot in a few days. However, an entire country of amateurs might accomplish even more.

In 2003 Patrick Meier, then a research fellow at Columbia University's Center for International Conflict Resolution, suggested that instantly updated maps could serve as potent tools for peacekeeping in conflict zones like the Horn of Africa. Aid workers, police, or soldiers could transmit reports of violence to a central office. There the incident would be layered onto a map showing the exact location of every trouble spot. By mapping all the incidents, peacekeepers could instantly identify the most dangerous hot spots, helping them respond quickly and with the right amount of force. Meier designed a basic interface for such a system, but his plan went nowhere. He lacked both money and tools. Existing geographic programs were too costly and complex; he needed software that was dirt cheap and dead simple.[11]

The tool kit Meier had longed for began to come together in 2008, in the aftermath of a bitterly disputed election in the East African nation of Kenya. Up to 1,300 people were killed in two months of rioting and ethnic violence. Appalled by the carnage, Kenyan attorney Ory Okolloh posted a message on her blog. "For the reconciliation process to occur at the local level the truth of what happened will first have to come out," she wrote. "Guys looking to do something— any techies out there willing to do a mashup of where the violence and destruction is occurring using Google Maps?"

One of those who answered the call was Erik Hersman. Born in the United States, raised in Sudan and Kenya, Hersman was working as a website consultant in Orlando. Okolloh's suggestion appealed to Hersman, a lifelong cartography buff. Moreover, Hersman believed that visualizing the violence in Kenya was the best way to capture the world's attention. "We realized that maps allowed us to make sense of things faster than if we just had lists of information."

Hersman teamed up with a pair of Kenyan expatriate computer programmers, David Kobia and Juliana Rotich. In two days they designed a website that would collect reports of violence from Kenyan citizens and display the information on a map. Informants with access to an Internet-connected computer could type in a report, including the name of the town, the date, and the time. Within minutes, the report would appear on the map.

The team called their new system Ushahidi—the Swahili word meaning "testimony." In the aftermath of the Kenyan crisis, a study conducted by the Kennedy School of Government at Harvard found that Ushahidi provided better information about the ongoing violence than any traditional media outlet.[12] The information collected by Ushahidi could help Kenyan citizens steer clear of danger zones. It ensured that Kenyan political leaders knew the real state of affairs on the streets and in the countryside. And it alerted foreign governments and the United Nations to the severity of the crisis. A later upgrade to Ushahidi enabled it to accept SMS text messages from cell phones, which are far more common in developing countries than personal computers and a lot more portable. With SMS eyewitnesses to a tragedy could file immediate reports from the scene of the event.

The Ushahidi team continued their upgrades, building the software into a complete system that could be used by any organization to generate real-time reports and constantly updated maps. Patrick Meier, who had dreamed of such a simple, powerful mapmaking tool back in 2003, joined the group as director of crisis mapping. Activists and aid organizations have since used Ushahidi to track dozens of events, ranging from natural disasters to hotly disputed elections in India and Afghanistan.

Perhaps Ushahidi's most notable success came in the aftermath of the horrific Haitian earthquake of 2010. More than 300,000 died; another 300,000 were injured, and 1 million people lost their homes. Aid agencies worldwide rushed disaster aid to the stricken island, but the devastation was so severe that existing maps of the country were often useless. Meier sent out a call for volunteers who would use Ushahidi to rapidly assemble new maps of the earthquake zone, based on information from news reports, quake survivors, and aid workers on the ground. Around the world volunteers formed "CrisisCamps," ad hoc gatherings that sorted through thousands of incoming requests for help. Haitian telecom companies quickly set up a special number to let people send text messages to CrisisCamps and broadcast the number over local radio and TV stations. Because 40 percent of Haitians own wireless phones, the CrisisCamp volunteers were soon receiving thousands of cries for help.[13]

Incoming reports were sent to teams of workers who summarized the information and linked it to the correct location on a map. Haitian expatriates who spoke the language were recruited to translate many of the messages; in other cases volunteers used software to render the messages into English. After translation the messages were scoured for street names and descriptions of landmarks, so the report could be added to the map. It was a frustrating task, as much of Haiti lacks street signs and some places are identified by multiple names. Furthermore, the text messages contained spelling errors, forcing volunteers to guess at their meaning.

Yet the volunteers kept at it, goaded by the plaintive messages begging for food and medical help. In all the volunteers processed more than a hundred thousand text messages. Most were translated and added to the map within two minutes of being received, ensuring that relief workers got nearly instant news of people in need. In addition, a host of relief agencies relied on Ushahidi maps and reports during the crisis. Officials of the US Marine Corps and the Federal Emergency Management Agency said that the data provided by Ushahidi was the best available in the immediate aftermath of the quake.[14]

Since the Haiti disaster relief agencies and political activists around the world have used Ushahidi maps in aid of their efforts. Volunteer firefighters in Russia used it in 2010 to map the progress of wildfires; environmentalists in Louisiana built a Ushahidi map to track the effects of the *Deepwater Horizon* oil spill; in 2012 the popular website *The Huffington Post* teamed with Ushahidi to map the comments of American voters in the run-up to the presidential election.

With tools like Ushahidi, it becomes relatively simple to overlay human needs and concerns onto our maps of the world, especially since the maps themselves exist primarily as easily modified digital files. Yet despite appearances, these maps are ultimately controlled by the powerful agencies that create them—governments and large corporations like Google that have spent millions building their vast geographic databases.

Then again, the same inexpensive digital tools that enabled the creation of Ushahidi have made it possible for determined amateurs to map entire nations almost from scratch and create new maps that belong to everyone.

In 2004 a British software entrepreneur named Steve Coast wanted to include customized maps on his website. To his surprise he learned that the mapping data would come at a ruinous price.

If Coast had wanted to make a map of the United States, he could have freely downloaded information from the TIGER database and modified it to fit his needs. That is because TIGER, like most intellectual properties created by the US government, is in the public domain. Anyone can use it, free of charge. Not so in many other countries, including the United Kingdom. That country's national and regional maps are prepared by a government agency called the Ordnance Survey. Founded in the eighteenth century to provide geographic data for the British military, the Ordnance Survey produces maps as fine as any on earth. However, the raw data from which they are made are copyrighted, and the government generated millions in revenue each year by selling the data to mapmakers.[15]

Coast was outraged. British citizens supported the work of the Ordnance Survey with their taxes; why should they have to pay again for access to the survey's data? However, his complaints were unavailing. If he wanted Ordnance Survey maps, Coast would have to pay. Or make a map of his own. Coast realized that handheld GPS units included a feature that let a traveler record his movements. A user could copy that track data onto a computer and combine it with details jotted down with pencil and paper—the names of the streets, buildings, parks, or public monuments that he had passed during the journey. He would now have an exact map of his journey, including the precise location of every object recorded along the way. GPS devices and personal computers made the process so simple that anybody could do it.

Coast went to work. He created the software tools he would need, purchased a GPS unit—at the time GPS devices cost hundreds of dollars and were "larger than a brick"—attached it to a laptop, packed the setup into a backpack, and started pedaling his bike through the streets of central London. Bit by bit, his map of the area began to come together.

Coast did more than pedal his bike; he also peddled his idea to anybody who would listen. He called it OpenStreetMap, a campaign to create a new kind of map, drawn by volunteers and available to everybody at no charge. Furthermore, OSM would not stop at creating a new map of the United Kingdom; Coast decided to cover the entire planet. That meant finding help, and a lot of it. Even though the participants would not earn a dime for their efforts, Coast believed they would step forward. "I think I was young and naive," he said.[16] Coast set up an Internet mailing list for mapping enthusiasts; he created a wiki, a do-it-yourself database of expert knowledge on do-it-yourself geography. And he talked and talked, at technology conferences, software user-group meetings, and every two weeks at a local pub where he would spend hours gulping down beers and talking up the power of amateur cartography.

Little by little, a growing number of his fellow Brits joined Coast's cause. Many were inspired by the success of the free computer operating system Linux. Assembled by thousands of volunteers scattered all over the world, Linux was once mocked as Tinkertoy software for amateurs. Today it is used by the world's largest businesses and government agencies. They also found a model in *Wikipedia*, the free online encyclopedia composed by amateur experts writing in their spare time. In the same way OSM would be entirely free for anyone to use, download, or copy. Even a for-profit business could reproduce the maps without paying royalties. Users would merely be required to credit OpenStreetMap as the source. And if they made additions or improvements, they would be obligated to share the new data with other OSM users.

Gradually, an accurate OSM map of London took shape. Just as important, Coast's idea began to catch on with people in other countries where mapping data were monopolized by governments or simply with hobbyists who fancied making their own mark on a new map of the world. Contributors did not even need costly equipment like a GPS unit or laptop computer. Once an area's basic road grid had been mapped, volunteers armed with pencils and paper could stroll the area to fill in details. These low-tech mappers did their part by recording the correct spellings of street names and landmarks.

By 2007 about 16,000 people worldwide were making contributions to OSM. It was not enough, of course. By then Britain had largely been mapped; the Netherlands was well covered, and so were portions of North America and South Africa. Yet most of the world remained a blank.[17] Still, the effort continued, gaining speed as more volunteers came aboard.

OpenStreetMap's success in Britain pushed its government into rethinking its map copyright policy. In late 2007 the Ordnance Survey offered a new application programming interface to let users generate customized maps on their websites.[18] The British government went even further in 2010. While the Ordnance Survey retained its copyright on the mapping data, the agency began granting free access via the Internet to the great majority of its maps.[19]

It was a welcome change, but OSM kept rolling full speed ahead, with contributors on every continent. In many parts of the world, the project's maps were as good as Google's offerings, perhaps even better. One reason was a boost from the commercial search service Yahoo!, which offered its own mapping service that included satellite and aerial photos. Yahoo! agreed to let OSM volunteers use its satellite images at no cost as a guide in drawing up their own maps. Access to these images sharply accelerated the work of the OSM volunteers. And the mapping campaign attracted a hoard of new recruits. By early 2013 1 million people worldwide had signed up as registered users of OpenStreetMap; of that number perhaps 300,000 contributed fresh data to the maps.[20]

In 2007 an executive at AOL's MapQuest said he was impressed by the dedication of the OSM volunteers, but added that his company would keep buying its mapping data from well-established commercial sources. These included Tele Atlas, a Dutch mapping company that has since been acquired by TomTom, a maker of GPS navigation units for cars, and Navteq, a Chicago mapping firm now owned by cell phone maker Nokia.[21]

By 2010, however, MapQuest officials were singing a different song. That year the company launched a new service in Britain that displayed mapping data from OSM. In addition, MapQuest's parent company, AOL, provided a $1 million grant to help the organization upgrade its maps of the United States.[22] Since then MapQuest has introduced OSM-based maps of many other regions of the world, including most of Europe, Japan, Canada, and the United States. That same year the OSM movement got a boost from one of Google's biggest rivals, Microsoft. The Seattle software company hired Coast to manage its own mapping service, Bing Maps. It also began donating Bing's library of satellite and aerial images to assist OSM volunteers in building better maps.

None of this posed much of a threat to Google's growing dominance of online mapping. It took a policy shift by Google itself to do that. In December 2011 the giant company announced that it would no longer offer free access to its API to major websites with large numbers of users. Hundreds of thousands of websites large and small used the API, and sites with fewer than twenty-five thousand visitors per day could keep on using the service at no charge. High-traffic sites would thenceforth be charged $4 for every one thousand visitors over the twenty-five-thousand limit.[23]

Google executives may have thought they had found a painless way to boost their revenues. Yet some of Google Maps' biggest commercial users were not so keen on the idea, especially when they knew that OSM's coverage of the United States was by 2012 good enough to serve as a substitute. And as a matter of contract and principle, OSM's maps would always be free.

Suddenly, OSM had a host of powerful new friends. Apple, for instance. Apple had long used Google's mapping service on its popular iPhone, but the two companies fell out over Google's rival smartphone software, Android. Apple ousted Google's maps and developed a new mapping product of its own. Apple's maps were obtained from a variety of sources, including commercial outfits like TomTom. However, Apple also used data from OSM. The social networking site

Foursquare, which helps people keep track of their friends' locations, built OSM maps into its software. And the online encyclopedia *Wikipedia* also adopted OSM maps in its smartphone apps.[24]

Alarmed by the defections to OSM, Google dialed back its price increase. "While the Maps API remains free for the vast majority of sites, some developers were worried about the potential costs," wrote product manager Thor Mitchell in a posting on the company's Geo Developers Blog. Google dropped the price from $4 to 50 cents per one thousand users.[25] The decision may have prevented further customer losses. Still, Google's careless handling of the matter had conferred new prestige upon a once obscure competitor. Its small but potent stable of major clients left no doubt that OSM was for real.

By this point even Google had long realized that Coast was on to something. In 2008 the company began its own effort to recruit amateur mapmakers with the release of Google Map Maker, a tool that let anybody modify and correct Google's online maps. Google had begun crowdsourcing its maps out of sheer necessity. As of 2008 the company still offered maps of just 22 countries and 20 million kilometers of roads—about 12.4 million miles. That sounds like a lot, but it is a mere fraction of the mappable earth. By 2012 Google Maps covered 187 countries and 42 million kilometers of roads, or 26 million miles. Much of the increase came as Google licensed map data from government agencies in countries like Russia and China. However, in Africa and the Middle East, most of the upgrades came from volunteers armed with Google Map Maker. "The big growth has actually come from user-contributed mapping—actually asking the local experts to help us out," said Ed Parsons, Google geospatial technologist. For example, the Moroccan city of Casablanca was nearly blank in 2008 because the government could not provide Google with an adequate map. Instead, city residents did the job on their own, in their spare time and with no pay.[26]

The citizens of Casablanca and other regions profit by having better maps of their environs; Google profits by selling commercial access to these improved maps. And the company does not have to share the wealth. Google asserts tight copyright control over all its maps, including any that have been generated entirely or in part by user contributions. Google is free to profit from its mapping service, but need not share a dime with the part-time cartographers who have worked so hard to improve them. By contrast, all OSM maps are published under a broad copyright. Any individual or business can copy, modify, or reuse OpenStreetMap's products to their heart's content. They are merely required to post a notice giving credit to OpenStreetMap and its contributors.

Few seem to mind Google's profiteering, as long as it results in better maps. The success of OpenStreetMap has provided healthy competition, as well as fodder for good-natured disputation. Which model produces better maps—Google's lavishly funded commercial operation or OpenStreetMap's legions of untrained part-time amateurs?

With its ample finances and direct access to the latest, sharpest satellite imagery, Google's got an overall edge in quality that OpenStreetMap will not soon match. Yet for one of its most valuable features, Google Maps relies on the kindness of strangers.

When you first activate a new Android smartphone, you'll see a screen asking for permission to switch on the phone's location tracking features. Hardly anyone refuses; the GPS and Wi-Fi location services enable the phone to provide users with turn-by-turn driving directions that most of us love. In addition, many popular apps rely on the location features; for instance, a banking app uses them to point you to the nearest ATM.

But even when you're not consciously using the location system, it's constantly at work, transmitting your whereabouts and your speed and direction of movement to Google's mapping servers. Google says there's no threat to privacy here. It neither knows nor cares who owns the phone; it merely knows that a given Android device is moving, say, north on Main Street at twenty miles per hour. And it gathers up the same kind of information from hundreds of millions of Android phones.[27]

With this information, the company can add remarkably accurate highway traffic data to their on-board maps. As a user drives down a clotted highway at rush hour, her location and motion are compared with data from surrounding vehicles. The more Android-equipped drivers on the road that day, the more precise the measurement of traffic. With hundreds of millions of smartphones in use, the traffic tracking software has plenty of data to work with. And that information is overlaid onto a Google map of local roads. Free-flowing roads are tinted green; more sluggish regions show up in yellow, and red means real trouble.

The resulting traffic maps are far more comprehensive than the road reports from drive-time radio DJs. Old-school traffic reports relied on data collected by state and local highway departments. With their limited resources, these agencies could track conditions only on major highways and arterial streets. Google's traffic maps, based on "crowdsourced" data from millions of vehicles, suffer from no such limitation. Users can get a traffic report on any street that attracts a sufficient number of Android-equipped drivers.

In mid-2013 the company expanded its commitment to crowdsourced cartography with one of the biggest acquisitions in its history, the $966 million deal to purchase an Israeli company called Waze. Founded in 2007, Waze offered a software app for iPhones, Androids, and other mobile devices that monitors traffic conditions in much the same way as Google Maps. But whereas Google relied on Android users merely as passive traffic monitors, Waze put them to work. By touching on-screen icons, "Wazers" can inform nearby drivers of a traffic accident, road construction, a barely visible speed trap. A driver or passenger can also transmit a text message, like "Big pothole on Fifth and Elm," and even include a photo.

By the end of 2012, Waze had signed up 36 million users in one hundred countries. Between them, they had shared 90 million reports on road repairs, police speed traps, traffic accidents, or garden-variety gridlock. As with most crowdsourced projects, only a tiny subset of users actively participate in making the service better. But though few in number, they do a prodigious amount of work. Just 65,000 users deliberately edit the maps, but these few users made a half-billion edits in 2012.[28]

By mid-2013 Waze had become one of the most coveted tech companies on earth. Several newspapers reported Apple had made a bid—a report that Apple denied. Then it was said that a billion-dollar acquisition by social media titan Facebook was virtually a done deal. But Google snatched away the prize, not by offering a bigger payout but by promising to allow the Waze development team to stay put in Israel, after Facebook had insisted they relocate to Silicon Valley.[29]

Waze continues as an independent mapping service. But its active user updates are now being blended into Google Maps. Along with the image of a red-tinted roadway, the maps may now display reports from Waze users, describing the three-car wreck that created the gridlock. At once, the resulting five-mile backup is no longer just an infuriating fact; it's now a narrative, a story of human misfortune, and for that reason a little easier to bear.

Even as we automate our maps, the end of cartography remains the desire to know where we humans stand on the earth, and how and why we got there. It's a big question that's answered in countless small places—intersections and railroad crossings, open-air markets and city blocks. And these days, the answers often come not from experts, but from diligent amateurs with open eyes.

Checking In

THESE DAYS NEW SMARTPHONE APPS ALL SEEM TO WANT THE SAME thing from us—our latitude and longitude. Whether it is serious or frivolous, whether it is a game or a social networking service, a digital cookbook or a video editing program, when you launch a new app for the first time, it will probably ask for permission "to use your current location." In a banal, nonthreatening way, you have been asked to reveal where you are, not only at that moment but throughout your day—and likely the rest of your life. It is an unsettling request, yet the great majority of smartphone owners, heedless of their privacy, willingly press "OK." At that moment, they have "checked in."

The term was first made popular by the social networking app Foursquare, whose users digitally inform their friends that they have arrived at a favorite restaurant, a fashionable nightclub, or the JetBlue terminal at New York's JFK Airport. Yet for all the buzz it has generated, only about 30 million people check in with Foursquare. Even so, according to a 2012 report from the Pew Research Center's Internet and American Life Project, three-quarters of America's smartphone owners use their devices to retrieve information related to their location—driving directions, dining suggestions, weather updates, the nearest ATM. The companies that deliver the answers must first know the phone's location. Most of us happily comply. In exchange for the information we need and the bargains we crave, most of us check in.

Early prophets of the Internet declared that cyberspace would make distance and location irrelevant. On the Internet nobody knows you are in Denver, or in Denmark, and even if people happened to know your location, it would not matter. Given a high-end computer and a high-speed broadband link, anyone anywhere could rival the mightiest global corporations and challenge the most despotic governments. Former Grateful Dead lyricist John Perry Barlow famously sang the song of the locationless Internet in his utopian manifesto of 1996: "Ours is a world that is both everywhere and nowhere, but it is not where bodies live. . . . Your legal concepts of property, expression, identity, movement, and context do not apply to us. They are all based on matter, and there is no matter here."[1]

Of course, the Internet is, ultimately, based in the physical world. It relies on a vast network of cables and computers, all made of the usual atoms and molecules. It is run by warm, fleshy humans, each of whom occupies specific locations on the planet. And the same goes for the billions who use the network. We are all here—and the exact position of "here," for each and every one of us, can be defined by latitude, longitude, and altitude.

Alternatively, our location is identifiable by our Internet protocol addresses. Every Internet-connected device has an IP address, a set of numbers that tells all other machines on the network how to contact your machine and pass on e-mail messages, Web pages, or Twitter tweets. Some devices, such as server computers at big companies, have permanent or "static" IP addresses. Most of the hundreds of millions of PCs, tablets, and cell phones are assigned temporary or "dynamic" addresses. That means that your smartphone's address might change from day to day. Although it sounds like a disaster for location tracking, as though a targeted consumer changed his home address from New York to Topeka to Baton Rouge on a daily basis, in reality, it is not so bad. Standard network diagnostic programs can track data between any two addresses on the Internet. These tests show the IP addresses and physical locations of each network router along the way. Trace a message to the end, and you have a general idea of where it wound up.

For example, the computer on which I am writing this book is presently connected to the Internet using IP address 76.119.24.212. When I typed this number into a website that offers an IP address lookup service, it announced that my Internet service was at that moment being routed through a server owned by Comcast and located somewhere in Quincy, Massachusetts. Right down the street, no doubt. Yet such a description needs further refinement, Quincy being a town of about ninety-two thousand. Computer scientists have found ways of running more advanced IP address tests that can narrow my location down to within a half mile or so. Yet even a rough estimate of your physical address is good enough for many marketers. Notice that the ads you see in your Internet browser often feature businesses based a few miles or a few blocks away. Travel to a different city, and you will see ads tailored to your new location. However, stay put and visit an overseas website. You will still get ads for American products, services, or stores, because the advertising server knows from your IP address that you are still on this side of the pond.[2]

Making Internet devices more portable has only enhanced the significance of location. Of course, the address of a housebound Web surfer is valuable information, particularly to those with something to sell. A thrifty marketer can avoid wasting the company's money trying to sell cheap malt liquor to someone with a home on Boston's Beacon Hill. However, the precise location of a smartphone owner cruising along Chicago's Lake Shore Drive is vastly more interesting. A vast matrix of information surrounds each of us, all the time. Yet when we start to move, the relevance and value of that information are constantly altered as we roll across the map. For example, a sale on handbags at a Michigan Avenue boutique might be mildly appealing to a woman miles away in Hyde Park, on Chicago's South Side. But let her phone notify her of the sale when she is walking past the store, and the bargain might prove irresistible.

When the woman gets into her car for the drive home, her phone takes a particular interest in her movements. Second by second, the phone's GPS chip measures her speed and direction of movement, as well as her location. Then it checks in, sharing the data with a distant map server. All around her thousands of other phones are doing the same. The aggregated result lets the woman's phone draw her a map that displays the viscosity of rush-hour traffic and urges her to consider an alternate route.

We roam our cities with navigational aids of magical precision. Most of the time we do not need them; it is a trip to work, to the supermarket, to church. Even then our apps and the people who created them want to know where we are.

Whether moving or standing still, the easiest way to find someone's location is simply to ask. An online retailer like Amazon.com, for instance, knows where you live because it must ship the merchandise to your door. This simple method of location will never go out of fashion. Early mobile advertisers took the same tack. In 2000 an Internet startup called Vindigo started providing cell phone users with restaurant reviews and shopping suggestions. No fancy location technology was involved. A customer looked up the information by typing in his current address. Vindigo's servers responded with listings of nearby shopping opportunities.

It was the early days of mobile location-based services, and first movers in the market often had a limited life span. Vindigo was acquired by a Japanese firm in 2004, and the new owners gave up on the business four years later. It would have been an unremarkable story about yet another tech start-up going under, except for the eventual career path of a onetime Vindigo employee named Dennis Crowley.

Crowley was laid off from the company in 2001 during the dot-com bust. After a few months of unemployment, he decamped to the Tisch School of the Arts at New York University where he entered the school's graduate program in interactive telecommunications. There the naturally gregarious Crowley and fellow graduate student Alex Rainert came up with a way to build on the original Vindigo concept. Instead of sending shopping or dining suggestions to users who entered their locations, why not put them in touch with other users who happened to be nearby? Such a service would make it easy for people to keep up with their current friends and quickly make a bunch of new ones.

By 2004 Crowley and Rainert had turned their idea into a product—a free mobile phone service called Dodgeball. The service worked on even the simplest cell phones because it relied on SMS text messaging. A user would sign up for the service at the Dodgeball website and create a list of friends who were also subscribers. If the user decided to go to a Boston Red Sox game, he could signal his friends with a text message featuring the familiar "at" sign—"@Fenway Park." Dodgeball's server computers would relay this message to the user's friends. If those friends had other friends who used Dodgeball, and if these "friends of friends" were within ten blocks of Fenway Park, they would receive the message too. This feature expanded the user's social circle, making it easier to strike up new friendships with the like-minded.

Within months of its launch, Dodgeball had acquired thousands of members in several major US cities. Yet its growth was hardly explosive. Dodgeball's use of text messaging, for one, was problematic. The technology was intensely popular in much of the world and would surge a few years later in the United States, but in 2004 most Americans still were not texting. Furthermore, a bar-hopping Dodgeball user would have to remember to send a fresh text update every time she shifted to a fresh watering hole.

Still, the company caught the attention of Google, which was already looking for ways to expand into social networking. Google acquired Dodgeball in May 2005 for an undisclosed sum; one report estimated a $30 million price tag. Crowley and Rainert stayed on to manage the operation, but not for long. Despite Google's investment the company seemed curiously uninterested in its prize. In April 2007, less than two years after selling Dodgeball to Google, the two founders left. "It's no real secret that Google wasn't supporting Dodgeball the way we expected," said Crowley. "The whole experience was incredibly frustrating for us."[3] Google kept Dodgeball on life support until early 2009 and then finally pulled the plug. Google's senior vice president of engineering Jeff Huber said the concept never caught on. "Maybe it worked in Manhattan," he said. "It didn't fly in Chicago, or St. Louis, or Denver, or the rest of the world."[4]

One reason, perhaps, for Dodgeball's troubles was the relatively primitive technology upon which it relied. Designed in the presmartphone era, the service compelled users to manually enter their locations. In 2009 Google supplanted Dodgeball with Latitude, a program for its Android smartphones that automatically tracked the phone's location and shared the information with the user's friends and family members.

In the five years since Dodgeball had made its debut, cell phones had become adept at figuring out where they were. The simplest wireless phone is a primitive homing beacon, announcing its presence to every nearby cell on the network. It had long been possible to get a rough estimate of a phone's location by identifying a cell that was communicating with it. In 1999 the US company Spotcast Communications used this crude but serviceable method to deliver advertisements to phone users in Hong Kong. Customers received discounts for putting up with a ten-second audio ad each time they placed a call. Spotcast would identify the cell site being used by the phone and then beam an ad tailored to the caller's approximate location. It was a rough approximation, as the phone could be anywhere within a mile or two of the cell site.

By 2009 mobile phones could be tracked with far greater precision. In theory you could use IP addresses to figure out the general location of a phone, as you can with a desktop computer. In addition, there are other more effective methods, thanks to the Federal Communications Commission. As mentioned in Chapter 5, the FCC in 1996 was worried that first responders could not locate people in danger who had used cell phones to call for help and ultimately issued regulations requiring that the cell phone carriers find a way to locate the users of their phones. The carriers came up with three solutions. The most exact requires adding a GPS chip to each phone, making it possible to pin down a caller's location to within 30 feet. Japan's cell phone carrier NTT DoCoMo had pioneered the effort, introducing the first GPS-equipped cell phones in 2000. Profit, rather than public safety, had been the goal. For an extra four dollars per month plus ten cents for each use, DoCoMo subscribers could punch up maps to guide their travels.

Adding GPS to a phone can be a serious drain on battery life, especially since it can take a minute or more to get a fix on the necessary four satellites. Moreover, a GPS device is not much use indoors, and even outside the signal could be blocked by trees or tall buildings. Still, some cellular carriers, like Verizon Wireless, immediately decided GPS was the best way to comply with the FCC mandate and began building the chips in all their phones.

These days virtually all smartphones and many cheaper phones have GPS built in. In the early days of the FCC mandate, however, some companies, like AT&T, wanted a simpler solution to the location problem. They stuck with the homing-beacon concept but sharply improved upon it using "time difference of arrival." In this method phones exchange signals with two or more of the network's cells. Each cell is equipped with a computer that calculates how long it takes for a phone to reply to an outgoing signal. The result reveals the distance to the phone. In addition, the cells calculate the direction from which the call came. Time difference of arrival can pin down a phone's location to within about 150 feet. Although considerably less precise than GPS, it is still adequate for most emergencies. Furthermore, it works with even the cheapest cell phone, as the device need not include a GPS chip. There is no extra burden on the phone's battery, because all the work is done by the phone company's computers.

Then came still the Wi-Fi technique conceived by the founders of Skyhook, introduced in Chapter 6. Companies like Apple and Google have imitated the concept and mapped the locations of millions of Wi-Fi hot spots, each of them transmitting a unique digital code. If the phone is within range of a few such hot spots, it can nail down its position to within 100 feet or so. Wi-Fi mapping is not much use in sparsely populated areas with few wireless hot spots to be mapped.

However, most Americans live in cities, making a Wi-Fi location service a generally reliable alternative to GPS.

Whatever the pros and cons of each location method, they produce formidably accurate results, especially when used in combination. Activate the location feature on an Apple or Android phone, and multiple tools may come into play, depending on which will produce the best results under the circumstances. Go someplace where there is no GPS signal, and the phone will try for a Wi-Fi fix instead. Google's decision to dispense with Dodgeball was thus unquestionably correct. Their replacement service, Latitude, did not need the user to type an address or the name of the nearest landmark. The phone already knew exactly where it was. With a flick of the touch screen, that information could be shared with others.

Though embittered by his Google experience, Dennis Crowley knew that the death of the original Dodgeball was no great loss. "It was really pretty lousy," he would later admit.[5] Although he left Google to join a company that created games for mobile devices, he clung to the idea of creating a social location program that exploited location-savvy devices like Apple's iPhone.

The result was Foursquare, jointly created by Crowley and Indian-born software engineer Naveen Selvadurai. With Foursquare there was no need for the user to punch his location into the phone. The phone already knew its own latitude and longitude, and Foursquare's database contained thousands of relevant points of interest—restaurants, bars, nightclubs. The software app displayed a listing of those nearby. A user seeking to round up friends for a night of revelry could simply select a likely-looking venue and "check in" with a finger tap. All her Foursquare drinking buddies now knew exactly where to find her.

It was a handy and amusing app for the social butterfly. And the phone's self-locating technology made using Foursquare practically painless. Yet people had plenty of other ways to keep in touch—using the phone to place a call, for instance. Foursquare would catch on only if it could change the habitual behavior of millions, by convincing them that the app was a better way to stay in touch. So Crowley and Selvadurai made a game of it. A Foursquare user who filed the most frequent check-ins at a particular location was declared the "mayor" of the place. The reward? At first a little cartoon badge and bragging rights among other habitués of the same location. Later, Foursquare worked with local merchants to offer discounts and special offers in exchange for check-ins.

Within two years of launching, Foursquare had acquired more than 6 million registered users, who were checking into the service 1.5 million times per day. Early on much of the activity was driven by the competitive urge to become the mayor of a favorite dining spot, public landmark, or subway station. In Philadelphia in 2010, for instance, Internet developer Andrew Miguelez fought ferociously to hang on to the mayorship of Penn's Landing, a well-known waterfront landmark. "I was in a heated battle with another daily visitor," said Miguelez. "The title of mayor bounded back and forth between us every couple of days as one of us would check in earlier than the other, or would check in on a weekend."[6] Competition is occasionally so intense that users cheat by checking into a location even when they are nowhere near it. Foursquare had to deploy software to spot and cancel fraudulent check-ins.[7]

By the beginning of 2013, Foursquare had attracted 30 million subscribers, with half of them having signed up in the previous year. Since its founding users have posted more than 3 billion check-ins. Although each of these may be valuable to a subscriber's curious friends, they may someday prove immensely profitable to Foursquare. Taken together these check-ins enable the company to track each user's movements, not only through the physical world but through the economy. The user who posts regular check-ins is leaving a series of bread crumbs that can enable marketers to figure out not only which retailers he frequents, but also which other nearby businesses might appeal to him, if only he knew about them.

Retailers can learn still more by aggregating the check-in data of millions. For example, they might discover that Foursquare users who frequently check in at art museums also like to shop at high-end clothing stores. Now art-loving subscribers who launch the app might be greeted with an ad from Lord & Taylor or some other specialty clothing chain. Foursquare, which has earned hardly any money in its brief life, is betting that this kind of precisely directed advertising will someday bring in a plentiful flow of cash.[8] Foursquare's social networking rivals agree and moved quickly to enhance their location-logging features. The billion or so members of Facebook can now attach their exact location to their posts, and so can the half-billion users of the messaging service Twitter.

Plenty of other schemes can make localized advertising pay off. The simplest are familiar to us from our desktop computers. Armed with IP address data, ad networks quickly figure out a user's general location—New York City, Chicago, or Spokane—and display locally relevant banner ads in our browsers. The same techniques are used by mobile ad networks. Those free smartphone apps so many of us enjoy are often paid for with small banner ads, some of them featuring the wares of local merchants.

Yet leading-edge location marketing goes far beyond such simple tactics. Collect enough location data, and a company can learn about far more than its customers' whereabouts. Just ask David Petersen, chief executive officer of Sense Networks, a New York City company that analyzes location data from millions of people to understand their buying behavior. When a smartphone user downloads one of those apps that asks to keep track of his location, there is a good chance that the data will be shared anonymously with Sense Networks. The company keeps track of the unique digital ID number found on every cell phone; it does not need the user's name to know all about him. "There's a thousand different things you can tell about people from where they go," said Petersen. There are obvious marketing clues—for instance, where you regularly shop. "Based on where we've seen a phone in the past, we can tell who's a Walmart shopper, who's a Target shopper."[9] Sense Networks digs even deeper. For instance, how often do you go out of town? Where do you go when you do? Do you often visit resorts or other fun places, or do you usually end up at a factory or office park?

Each of these location data points, aggregated and analyzed from millions of smartphone users, enables Petersen's company to assign consumers to lifestyle categories—business travelers, for example, or college students or avid gamblers. When an advertiser needs to reach, say, a million mobile phone owners with a fanatical love of sports, Sense Networks has already tracked them down by keeping track of the places they go. The company does not know who they are, only the digital IDs of their phones, which will soon be on the receiving end of a series of Super Bowl ads or sports-apparel discount coupons.

With location-aware smartphones, advertisers can transcend the merely local. They can begin beaming us hyperlocal advertising, tailored not just to the city, but to a particular city block. The idea is called "geofencing," an unfortunate name choice that evokes the ankle bracelets sometimes worn by accused criminals under constant surveillance. The earliest such devices fenced in the user by transmitting a radio signal to a box connected to his home telephone line. If the suspect left the building, the radio signal would fade, and the box would place an automated phone call to the cops. With the addition of GPS and cellular technology, later versions of ankle bracelet technology allowed a greater measure of mobility. A judge might grant a criminal suspect permission to go to her job, her church, and her local supermarket, with each approved location plugged into the court's computer system. Data from the ankle-strapped GPS could confirm that the suspect was staying out of mischief or send a warning to police when she paid an unauthorized visit to the local dive bar.

Geofencing also has uses for the law abiding. A company called Life360 uses it to help parents keep tabs on their kids. The service homes in on location data from a child's phone and sends a digital message whenever the kid arrives at home or at school—and whenever he leaves. Stroll off campus at ten in the morning, and the parents instantly know. As of late 2012, Life360 had signed up about 25 million users.[10]

When marketers build a geofence, they have no desire to restrict our movements. They want us out and about, constantly traveling past places where we can spend money. Far from building fences, they are stringing trip wires. The goal is to detect our close approach to a nearby business that is looking to make a sale, so the company can ping you with a text message urging you to buy. "I think we were probably almost the first to deploy it at scale, four years ago," said Alistair Goodman, chief executive officer of Placecast, a San Francisco company that has become the most prominent practitioner of geofencing. "We saw that the physical world and the digital world were going to collide." Goodman founded Placecast in 2005, convinced that the FCC's 911 mandate, and the plummeting prices of phones, would bring location-based selling to the masses. "When my 80-year-old father told me he had to have a cell phone," Goodman said, "that's when I knew it was happening."[11]

Retailers like Starbucks, Kmart, and the Subway restaurant chain deliver ads through Placecast's ShopAlerts service. The company also has alliances with US cellular carrier AT&T and the European phone company O2. In all Placecast delivers geofenced ads to 10 million phone users in the United States and Britain. Each potential customer wants to receive the ads; Placecast works on an opt-in basis. For example, a sandwich lover might ask to get a text message when he is within a block of a Subway store.

Goodman realizes that nobody wants a constant stream of text messages. With coffee bars and fast-food restaurants on every block, life in a geofenced world could soon become intolerable. So Placecast practices a policy of "frequency capping." Customers generally get no more than five messages a week, even if many other attractive deals come within range. Subscribers do not need to own a GPS-equipped smartphone, either. "GPS can get you to within 50 feet or even closer" to a local store, said Goodman, but "you don't actually need that level of precision." Even a crude location fix obtained by triangulating to the nearest cell towers is good enough for a geofence. Placecast has erected geofences around 262,000 locations in the United States and United Kingdom. The company claims that one out of two consumers who have subscribed to the ShopAlert service has visited a merchant after being notified of a special offer, and 22 percent end up buying something. Of those who spend money, half had not planned to buy anything until their phone suggested it.

As with other forms of location-based advertising, Placecast's geofencing system is anonymized, to ensure that the company cannot identify the people it is tracking. Location data are saved for thirty days so the marketing experts can analyze the results of the campaign; after that it's tossed. Asif Khan, founder of the Location-Based Marketing Association, an industry trade group, says that Placecast and other such companies take great pains to protect sensitive data. But Khan argues that consumers are not all that worried; give them a bargain, and they'll let you draw a bull's-eye on them. "Nobody cares about their privacy, as long as they get what they want."

Yet geofencing is rarely used by advertisers. Our cell phone batteries get part of the blame. Geofencing requires constantly recalculating the phone's position, a habit that shortens battery life. "The battery-drain issue has been an issue for years," Goodman admits, adding that his company has developed software algorithms that pin down the phone's location more efficiently. Hardware makers have also tackled the problem. In February 2013 Broadcom, a major maker of the chips used in smartphones, introduced a new GPS chip that is designed to run at full throttle when the user needs turn-by-turn driving instructions. The same chip goes into battery-saving low-power mode when running in the background and watching for geofences.[12]

Even if geofencing becomes more energy efficient, it is still not a sound strategy for selling many consumer products. "We've found that it doesn't really work very well," says David Petersen of Sense Networks, because "we as humans don't really consume things spontaneously." At least, not very valuable things. Alert someone to a half-price sale on soap at a nearby store and he might pop in. But hardly anyone will pull off the highway and into the mall merely because his phone announces a half-price sale on flat-panel TVs. Even at the lower price, a good TV will cost hundreds of dollars. It is the sort of purchase people think about and plan for. As a result, says Petersen, TVs and pretty much every other big-ticket buy are off-limits for geofencing.

A more promising location-based strategy targets the last frontier of navigation—inner space. After a century of innovation in geotechnology, nearly every square foot of land on the planet has been mapped. Step through the doors of a shopping mall or airport, however, and it is easy to get lost. GPS won't help; satellite signals rarely penetrate the walls. What shoppers need is a good map. Or better still, a good app, one that could display interior maps for thousands of likely destinations. It is a problem being tackled by dozens of companies, from mapping giants like Google to little start-ups with names like Wifarer and Point Inside. And already, their work is paying off. Visit a major airport, museum, or shopping mall in the United States, and there is a good chance that your smartphone can punch up a detailed map of its interior.

Google, for instance, has mapped more than ten thousand large structures worldwide and is urging real estate developers to supply their floor plans. The information is displayed in the standard Google Maps interface. Launch the company's map app and look up Chicago's O'Hare Airport. If you zoom in for a close-up of a passenger terminal, you will see the names of the shops lining each corridor. Point Inside, based in Bellevue, Washington, and founded in 2009, delivers a much more detailed product. It is has produced maps of hundreds of major venues, all of them easily accessible through a free app. Apart from displaying a simple floor plan, the Point Inside app is searchable. Punch in "Wolfgang Puck," and you get a listing of the dining entrepreneur's four restaurants inside O'Hare, along with hours of operation, a description of the cuisine, and of course a map.

Shoppers at Walgreens drugstores may never again have to ask which shelf holds the allergy medications. Instead, there is a smartphone app from Aisle411, a St. Louis–based company that has mapped all of Walgreens' seventy-nine hundred US retail stores. Even before a user gets to the store, he can launch the Aisle411 app. Using the phone's location features, the app displays a list of the nearest Walgreens stores. The shopper can pick his favorite and then peck in a search for allergy drugs. Up comes a simple map of the store with pointers to several locations: allergy medications for kids, travel-size containers for tourists, drugs especially for the eyes or the throat. A shopper in a hurry will know where to find the correct product before he crosses the threshold. The app also features a recipe planner.

Peck in an ingredient (fish, for instance) and the preferred method of cooking (grilling, perhaps), and it displays recipes scooped up from a number of online cooking sites. Then it offers to plug the complete ingredient list into its mapping software, showing you exactly where in the store you will find each item. This feature will prove quite handy if Aisle411 succeeds in its goal of signing up major supermarket chains.

The Aisle411 app has an indoor geofencing function as well. By measuring the strength and direction of Wi-Fi signals from several routers in the store, the app can calculate the user's position to within a few feet—close enough to know that he is strolling down the soft drink aisle. If there is a special on Coca-Cola that week, his phone will let him know just as the brown bottles come into view.

Another indoor navigation start-up, Wifarer, uses indoor Wi-Fi mapping to provide step-by-step guidance through malls. Available at a handful of locations, including the Prudential Center mall in Boston, Wifarer lets a visitor enter the name of a particular landmark inside the structure—a bookstore, for instance. The Wifarer app uses Wi-Fi triangulation to show the user's position on a map of the mall and then displays a dotted line leading to his destination.

Another innovative shopping app called Shopkick uses sound to steer shoppers through large retail stores like electronics seller Best Buy. It is high-frequency sound that human ears cannot detect, but comes through sweet and clear to a smartphone. Shoppers can earn discounts and other rewards by launching Shopkick as they enter the store. They are promised more rewards for visiting certain parts of the store to check out special deals on, say, cell phones or video games. Shopkick knows instantly if they comply; by analyzing the sound waves, the app knows each shopper's position inside the store to within a few feet.

It is not unlikely that in a few years, the interior of every major public venue will be mapped with the same sort of precision as our streets. No more getting lost on our way to the food court—surely a good thing. But our access to near-perfect navigation comes at a cost. We may know exactly where we are at all times, but others know as well, whether we like it or not.

On the Spot

IF YOU WENT LOOKING FOR SOMEONE TO DEFEND YOUR CIVIL LIBERTIES, you would probably opt for a character from a John Grisham novel—brilliant, articulate, and armed with an Ivy League law degree. Yet citizens who dread being kept under relentless surveillance by an intrusive government had to settle for a rather less admirable champion—Antoine Jones, a convicted drug dealer turned Washington, DC, nightclub owner.

In 2005 it looked like time for Jones to pack his toothbrush again. He and several associates were arrested in a combined city and federal drug raid that turned up more than a hundred kilograms of cocaine and $850,000 in cash. It was the biggest drug seizure in the city's history. And it had all been made possible by the cunning application of GPS technology.[1] Yet in due time the very locational technology that had promised an open-and-shut case for the prosecution would lead them to defeat, establishing a new legal precedent in defense of a citizen's right to be left alone. There are limits to how far the government can go in using advanced location techniques to keep tabs on us.

But those limits will have to be imposed by legislation, judicial fiat, or public outrage. They are no longer fixed by technology. As we have seen in the preceding chapters, mankind has essentially solved the problem of location. Thanks to cellular phones, Wi-Fi, global positioning system satellites, and dagger-sharp aerial and satellite photos of the planet, it is possible for the first time in human history for anybody to find out where he is, relative to anybody or anything else on the surface of the earth. Yet it is just as easy for government agencies and corporations to track us. The same smartphones that steer us to the mall constantly transmit our exact locations. And phones are being supplemented with other location methods that are even subtler, and often far more insidious.

For centuries people have been able to disappear. By moving to a new continent or a new city, a man could effectively vanish from the sight of his fellows. He could cast off the dead weight of past mistakes and create a new life for himself. Not anymore. With its Social Security numbers, credit cards, and driver's licenses, the twentieth century made it nearly impossible to live the anonymous life. Governments and businesses track our dollars, take note of our employers, memorize the addresses of every home we have ever lived in, and retain the vehicle identification numbers of every car we have ever driven.

At least we could take comfort in our ability to move around undetected. Not by plane, perhaps, but in our cars and trucks, on buses, or by foot. In motion at least, there was anonymity and freedom. No longer. In a world where our phones transmit our location whenever they are switched on; where traffic lights, toll booths, and even cruising police cars include sensors that identify every vehicle that comes into range; or where the simple ID tags worn by workers and students quietly broadcast radio messages revealing our movements, we can no longer move stealthily on our daily commutes, shopping trips, or walks.

Yet it seems that despite our intermittent grousing, locational privacy is something we do not seem to mind giving up. We have come to love the way our phones tell us the local news headlines and weather because they know exactly where we are, the way we can roll through highway toll plazas and barely slow down, because the radio transponders in our cars pay the toll. So what if the transponder also tells the state highway department that we drove east toward Boston or west toward Worcester at 9:35 a.m. on Monday? Do phone makers or advertisers or governments know more about our movements than they truly need to know? Perhaps, but there are so many benefits. And after all, we have nothing to hide.

"Geospatial data is analytic superfood," says Jeff Jonas, a high school dropout who has become one of the top researchers at IBM, because it is so remarkably useful. From an individual point of view, it helps us find our way to the best Mexican restaurant in town or to the nearest hospital emergency room. However, it is even more useful to any organization that can study in aggregate the moves that each of us makes. For instance, researchers have found that if you track someone's cell phone usage patterns over a three-month period, you will be able to predict where this person will be in the future with an accuracy of 93 percent.[2] A police force could use this predictive method to arrive at the scene of a future crime an hour ahead of the prime suspect. Yet they might also be able to scour millions of cell phone location records and thereby predict the movements of political dissidents. Our ability to constantly track each other's locations is "going to change our existing notions of privacy," says Jonas. "A surveillance society is not only inevitable and irreversible; it's worse. It's irresistible."[3]

Thanks to our mastery of location, we may never be truly invisible again. To preserve some shred of anonymity, we must look to laws; perhaps we can compel those who watch our movements to avert their eyes, at least occasionally. Or maybe technology can rescue us once more—this time with inventions that allow us to enjoy the benefits of location systems without flinging off the cloak that conceals our own movements.

For those seeking legal protection, the Jones case had a happy outcome, though perhaps not entirely satisfying. The Supreme Court's ruling still leaves questions about how far governments may go in tracking us. Moreover, criminal prosecutors are not the only ones who may want to know where we are going. As we saw in the previous chapter, there is money to be made from location data, and abuse of that information by corporations could be almost as harmful as the most irksome intrusions of an overzealous government.

In fact, most of the systems that track where we are going are owned and operated not by busybody bureaucrats or secret policemen, but by corporations. Our cell phone carriers must track our locations in order to stay in touch with us. However, the companies that create the software driving our smartphones, companies like Apple and Google and Microsoft, also build an ever-expanding history of our locations and movements and trade on this information for profit.

We are even tracked by the third-party software apps we install on our phones. A research report from February 2013 found that half of the fifty most popular apps for iPhones and Android smartphones automatically transmit information about the user's location.[4] Fire up the popular game Angry Birds, and game designer Rovio will use the phone's GPS and Wi-Fi to determine where you are. It is useful information for pinging you with targeted advertisements. Yet in 2012 researchers at the Massachusetts Institute of Technology found that when a user stops playing Angry Birds, it keeps right on running in the background, steadily broadcasting the player's whereabouts.[5]

Add to this the detailed personal information available from information brokers like Acxiom, which has files on about 190 million Americans and a total of a half-billion people worldwide. Acxiom knows the age, race, sex, weight, height, marital status, education level, even political leanings of millions of us, based on data culled from publicly available sources.[6] Anybody from the police to commercial enterprises can purchase this data and combine it with location tracking to create a comprehensive survey of our lives. As John Gilliom and Torin Monahan wrote in their book, *SuperVision,* "For most of us, surveillance comes not from a unitary state bent only on domination and control, but from a chaotic blend of government, media, work, friends, family, insurance companies, bankers and automated data processing systems."[7]

Still, it is the threat of constant overwatch by surveillance-happy governments that is most alarming. However much Google may know about you, it does not run any maximum-security prisons. That is the destiny that the US government had in mind for Jones when it began tracking him in 2005. That year a federal district court issued a warrant that allowed federal drug investigators to plant a GPS tracking device on a Jeep registered in the name of Jones's wife, which Jones himself routinely drove. The tracker recorded every movement of the vehicle to an accuracy of one hundred feet or less. All that data—two thousand pages of it—was transmitted to cops via the local cellular phone network during a four-week period. The location data left no doubt that Jones was a regular visitor to the drug stash.[8]

Armed with so much evidence, it is surprising that Jones's first trial, in 2007, resulted in a hung jury. It seems the jurors were troubled by the fact that Jones was never found in possession of drugs and was never seen at the drug house. All the government had was testimony from some codefendants and the GPS readouts. Not enough, said the first jury. But a second group of citizens was persuaded. Jones was found guilty of conspiracy to distribute cocaine and sentenced to life imprisonment.

That might have been the end of the story, except that Jones appealed to the federal court of appeals for the District of Columbia and won a unanimous ruling from a three-judge panel that threw out the conviction. It seems that the warrant to place the GPS device was valid for ten days, but the government did not place the device until eleven days later. Then they had kept it going for twenty-eight days. To the appeals court judges, these blunders had transformed a precisely targeted surveillance mission into an open-ended fishing expedition, thereby violating the Fourth Amendment's ban on unreasonable searches and seizures.

The federal government appealed the ruling to the US Supreme Court, declaring that the lack of a valid warrant was no barrier to GPS surveillance. After all, the GPS sensor merely identified the location of a vehicle traveling on public roads; a person cannot very well argue that his public comings and goings are a private matter. By that line of reasoning, a cop would have no right to tail a suspicious person prowling the streets, without first asking the person's leave or getting a warrant. "Any individual who moves on public roadways knows that his movements can be readily observed," the Obama administration argued in its appeal brief. Jones, the government said, "had to be aware that any neighbor could have observed his frequent visits to his Fort Washington stash house." And if the neighbors could have sussed it out, there was nothing wrong with the government doing the same, with a little high-tech help.[9]

The High Court was not convinced. In January 2012 it ruled unanimously that the placement of the GPS device on the Jeep amounted to a search of the type described by the Fourth Amendment. As such, the GPS surveillance should not have been conducted except under the authority of a proper search warrant. As the original warrant had expired, none of the evidence obtained thereby could be used in prosecuting Mr. Jones, and the guilty verdict was voided.

Civil libertarians were not altogether pleased with the majority opinion, composed by Associate Justice Antonin Scalia, and joined by Chief Justice John Roberts and Associate Justices Anthony Kennedy, Clarence Thomas, and Sonya Sotomayor. Scalia and his four colleagues declared that the planting of the GPS device became an illegal search as soon as the investigators laid hands on Jones's vehicle. The car was Jones's property, and intruding upon it to install a tracking device violated Jones's Fourth Amendment rights.

Four justices, including Sotomayor, who had signed off on Scalia's opinion, wanted to go much further. Led by Associate Justice Samuel Alito, they believed that the mere fact that Jones's movements had been electronically tracked without a warrant for nearly a month was unconstitutional. Alito based his argument on a famous 1967 case, *Katz v. US*, in which a bookie successfully appealed his federal conviction for running an illegal gambling operation. Katz was caught because the FBI had tapped the pay phone he used to place his wagers. Yet the court ruled seven to one that the conviction was void because the FBI had failed to obtain a warrant. The US government argued that because the calls had been placed from a public phone booth, no warrant was necessary. The court disagreed, saying that even when using a public phone, Katz had "a reasonable expectation of privacy"—the right to assume that he could speak freely without being monitored by government agents.

Katz established a new standard for investigators. Whenever a citizen has good reason to believe his actions are private—when he is at home, or in the men's room, or in a phone booth—the government cannot conduct surveillance without a warrant. In the *Jones* case, Alito and four of his colleagues had wanted to apply this tougher standard. Any long-term tracking of a citizen's movements must require a warrant, they said, because people have a reasonable expectation that their movements *over time* are a private matter. Some cars already have built-in GPS trackers to assist in recovery if the vehicle is stolen. Such a system could be activated by government agents and used to track a vehicle for days or weeks, without physically trespassing on the suspect's property. Under Scalia's ruling, such a tactic would be permissible because it would not involve an act of trespass. Yet for Alito the surveillance alone was reason enough to overturn the verdict. "Relatively short-term monitoring of a person's movements on public streets accords with expectations of privacy that our society has recognized as reasonable," wrote Alito. In other words, an undercover cop does not need a warrant to follow your public movements for a few hours or days. "But the use of longer-term GPS monitoring in investigations of most offenses impinges on expectations of privacy," Alito added.

The ruling voided Jones's conviction. The US Justice Department was free to retry the case, using whatever untainted evidence remained. It did exactly that, by tapping a private-sector surveillance network. Armed with a subpoena, the government obtained five months' worth of Jones's cell phone records, including the number dialed, the time of the call, how long it lasted, and, perhaps most important, the location of the phone. Thanks to the aforementioned *Katz* decision, the police must have a court-issued warrant to listen to someone's phone calls. And to get such a warrant, the police must show "probable cause"—specific evidence that indicates the likely guilt of a suspect. However, a subpoena for records of a person's phone usage is generally much easier to get. Investigators need tell the judge only that they believe the phone records contain information "relevant and material to an ongoing criminal investigation."[10]

The police have been taking full advantage of this easier standard. In 2012, at the behest of Massachusetts US representative Edward Markey, the nation's cell phone companies revealed that they had handed over customer records to law enforcement agencies 1.3 million times in 2011. Put another way, some federal, state, or local police agency was requesting someone's cell phone records nearly thirty-six hundred times every day.

With several months or years of cell phone location data, police can track a person's movements with great precision. It is even possible for investigators to be waiting for the subject before he arrives at his next appointment. As we have already seen, three months' worth of location data can let researchers predict a person's next move with amazing accuracy. And like the Internet marketers we met in Chapter 9, investigators can combine the location data with data stored in police files or purchased from commercial data brokers to reconstruct nearly every aspect of a person's life—including the likely headquarters of his illegal drug business.

The prosecutors had not used Jones's cell phone records in the first trial because tracking the Jeep via GPS was much more precise. Still, the phone logs were accurate enough to show that Jones was frequently in the vicinity of the drug den. Combined with video footage from local surveillance cameras showing Jones's Jeep visiting the place and testimony from the aforementioned informants, the government believed it could still win a conviction. After all, the prosecutors could call upon all the resources of the Justice Department; Jones was reduced to acting as his own attorney. Yet despite such efforts, in March 2013 Jones won another mistrial, when the jury declared itself deadlocked. Jones remains in custody, and federal prosecutors insist they will try again.[11]

In some cases the police do not have to rely on customer records from cell phone companies. Instead, they set up their own cellular system and trick people into using it. A technology called Stingray, employed by police agencies since the 1990s, creates fake cell towers that can help investigators lock onto the location of any wireless phone in the vicinity. In principle, the concept is simple. Your cell phone announces its location with every signal it sends, transmitting a signal to every nearby cell tower owned by your phone company. A Stingray system is a portable device, easily mounted in a car. It intercepts the incoming signals of nearby phones and sends an acknowledgment that tricks the phone into believing that the Stingray is a legitimate cell site. The Stingray records basic information about the phone, including its unique digital ID code, and then measures the strength of the signal coming from the target phone. Next, the car containing the Stingray moves to a different location, reconnects to the suspect's phone, and recalculates signal strength. By repeating this process at multiple locations, the Stingray can calculate the position of the phone to within a few yards. Obviously, the suspect's phone is not the only one that can respond to a Stingray; any other phone in the area could also be tracked. Law enforcement officials say that they simply discard this extraneous information, to protect the privacy of those not under investigation.

Stingrays have been in use for nearly twenty years, by the Federal Bureau of Investigation and other law enforcement agencies, including the police departments of Los Angeles, Miami, Fort Worth, and Gilbert, Arizona.[12] Still, the technology has only recently attracted public notice, thanks to Daniel David Rigmaiden, who in 2010 was charged by federal prosecutors with tax fraud. Rigmaiden was tracked down when investigators used a Stingray to locate his laptop computer, which was equipped with a data modem that connected to the cell phone network.[13]

Rigmaiden has argued that he should be granted access to complete technical details about how the Stingray works, in order to challenge the validity of the evidence against him. Rigmaiden believes that using a Stingray to track him down constitutes a search of the kind that requires a warrant—which the FBI did not have. The FBI does not want to reveal details of how Stingray works. And the agency insists that figuring out Rigmaiden's location by pinging his cellular data modem is not enough of a search to require a warrant, because there is no reasonable expectation of privacy about cell phone location data.[14] For now, the *Rigmaiden* case still languishes in the courts. Its eventual resolution should provide us with further guidance about how far the police can go in tracking the location of criminal suspects.

If the police cannot track our phones, they can still follow our cars and track us by reading our license plate numbers. If that sounds like a limited, low-tech approach to crime fighting, think again. Police forces throughout the United States track millions of vehicles with help from digital systems that record the license plates of passing cars. If you are a driver, your vehicle has probably been tracked this way. A 2009 study found that 37 percent of police agencies of more than one hundred officers were using license-plate recognition systems, and another one-third of departments were planning to get it.[15]

In an LPR system, police mount cameras along major roads, at intersections, or even on roving squad cars. These cameras give a clear view of the license plates on passing vehicles. The large bold letters and numbers are easily readable by a computer. The machine records the time at which the image was captured, and with help from a GPS receiver, it pinpoints the location of each scanned vehicle. In effect, LPR systems raise the same issues posed by other forms of constant surveillance. It is perfectly proper for the cops to note license plate numbers in the course of an investigation, and someone driving down a public street has no right to license plate privacy. However, with license plate cameras at every major intersection and every patrol car, drivers are effectively under constant police surveillance, whether they are criminal suspects or not. And all this captured data can be saved—and frequently is. That means that police can track your movements and locate your favorite hangouts without ever putting a tail on you personally. Instead, they can call up the license plate tracking database and see where you have driven for the past six months.

San Diego County in California has been running an LPR system since 2010 and has stored the results of more than 36 million scans, or an average of 14 for every car or truck in the county. One journalist who wrote about the system learned that his own car's plate had been scanned 24 times in thirteen months. Local police say that they have used the stored data to solve a horde of serious cases, ranging from hit-and-runs to homicides. "I've lost count of how many crimes we've solved because we've been able to go back and pinpoint locations," said Sergeant Scott Walters of the Escondido Police Department in San Diego County. While some police agencies keep the records for a year or two, San Diego County Sheriff's Department plans to store LPR files indefinitely. "Short of a criminal investigation, no one is accessing this data," said sheriff's deputy Lieutenant Glenn Giannantonio, "so there really is no need to dispose of it."

Privacy advocates worry that LPR systems will enable governments to store a permanent record of every driver's comings and goings. "That is extremely frightening," said David Loy, legal director of the American Civil Liberties Union of San Diego and Imperial Counties. "I don't think that's the world we want to live in." Loy and other civil libertarians favor laws that would set a ten-day limit on the storage of LPR data—time enough to aid investigation of recent crimes, while preventing random long-term monitoring of citizens.[16]

Surprisingly, whereas GPS tracking and mining of cell phone records have generated a legal and legislative backlash, LPR surveillance has created relatively little controversy, perhaps because few people know about it. Until somebody files a lawsuit or a proprivacy lawmaker in Congress files a bill to regulate the practice, police forces throughout America will keep right on using license plate scanners to record our movements, whether we like it or not.

While the surveillance efforts of federal agents were getting smacked down, the public schools of San Antonio, Texas, have fared much better in defending a surveillance technology that is far more common and arguably even more intrusive—radio-frequency identification. The RFID concept dates back to World War II and Robert Watson-Watt, the British radar pioneer we met in Chapter 2.

Watson-Watt realized that in a sky full of fast-moving military aircraft, it might be difficult to distinguish allies from enemies. So he devised a system called IFF—Identify Friend or Foe. A radio on the ground would broadcast a specific frequency. British aircraft would carry dedicated radios that would listen for this frequency and instantly respond with a signal confirming their identities. The IFF system was the forerunner of the transponders on today's commercial planes that identify the aircraft to ground controllers.

RFID is a modernized, miniaturized version of the same idea. A small microchip with attached radio antenna can be embedded into an ID card or glued onto a package. This device sits inert until it comes within range of a "reader" that broadcasts a radio signal. The incoming signal alerts the RFID chip, which responds by beaming out a serial number. This serial number has already been stored in a database, where it is associated with a particular object or person. So when the RFID reader at a supermarket warehouse picks up the serial number of an RFID chip attached to a box of spaghetti sauce, it notifies the warehouse's inventory computer that the sauce has arrived.

With RFID you do not need warehouse workers typing in the data as new shipments arrive. You just put RFID readers at each loading dock; the merchandise logs itself into inventory as it is brought into the warehouse and logs itself out when it is shipped out to stores. Retailers like Walmart who want to control labor costs depend quite heavily on RFID systems; so do manufacturers like airplane builders Boeing and Airbus, which use RFID tags to track vast inventories of spare parts.[17] RFID chips also simplify many everyday human transactions. In Boston commuters use cards with embedded RFID chips to pay their fares by simply waving them in front of a detector. Millions of people pay their highway tolls with an RFID-based "EZ Pass" that identifies the driver and sends him a bill each month. And many employers issue RFID-based identification cards to unlock the office door and confirm the worker is on the job.

Plainly, RFID enables location tracking at its most granular level, even when it was not deployed for that purpose. EZ Pass toll systems, for example, were designed to speed up traffic and reduce the need for human toll takers. However, every time an EZ Pass–equipped vehicle passes an electronic tollbooth, its passage is recorded in a database. Cops or attorneys armed with subpoenas routinely inspect these records, which could provide an alibi in a criminal inquiry or prove to a divorce lawyer that you were not working late at the office that night.[18] On an even more granular level, retailers have taken to attaching RFID tags to individual items in their stores as a precise tool for tracking inventories. Some civil libertarians fear that it is the first step in a process to track nearly every aspect of our lives—for example, RFID chips woven into clothing would constantly transmit our comings and goings.

RFID industry executives say they are not interested. Still, in their 2005 book, *Spychips,* Katherine Albrecht and Liz McIntyre uncovered documentation from US companies suggesting that they have at least been thinking about it. For instance, in 2001 IBM filed a patent on a way of using RFID tags to track people inside "shopping malls, airports, train stations, bus stations, elevators, trains, airplanes, rest rooms, sports arenas, libraries, theaters, museums, etc." IBM has not followed through on the idea, at least not yet.[19]

Some RFID applications are intended to keep track of the holders' whereabouts. Sometimes it is an imprecise, soft form of tracking. Use an RFID name badge to enter your office or clock in at the factory, and the company has a record that you showed up that day. Still, that probably does not mean your movements inside the building are being constantly monitored.

Most RFID tags are "passive." They are incapable of transmitting radio messages on their own because they do not include a battery to power up the chip. Instead, passive tags come to life only when they are near an RFID reader. As the chip approaches the reader, it scavenges power from the incoming radio signal—enough for it to answer back with the chip's ID code. Because the reader's radio signal is weak, passive RFID chips respond only when they are close to the reader—usually within a few yards or a couple of feet.[20] Plainly, such tags cannot constantly track your exact location in a large building. In most cases that is not necessary. Just put a reader in the doorway of every important room. Now a bank's security system knows when a particular worker enters the vice president's office or walks into the vault.

RFID systems can be made far more intrusive by deploying tag readers in every room, even in places where workers might expect some privacy. Back in 2004 a company called Woodward Laboratories introduced iHygiene, an RFID-equipped soap dispenser for the bathrooms of fast-food restaurants. The iHygiene dispenser would detect the RFID badge of any employee using the room. If that worker departed without putting soap on his hands, iHygiene would send a signal via Wi-Fi to the store manager's computer. Happily, the product never caught on.[21] Another version of RFID could prove even more intrusive. Add a small battery, and you can create an "active" tag that would transmit a signal for dozens of feet. A passive chip signals the wearer's location only when he is passing near a reader. With its much longer range, an active RFID chip can keep a worker constantly on the company radar.

Hospitals have led the way in using active RFID systems designed to constantly track the location of people and equipment. However, results have been decidedly mixed. RFID signals are often far less precise than advertised, so that tagged people and items cannot be exactly located. The batteries in active tags must be regularly replaced—a job often neglected. Many hospitals have found the technology much too costly. And there is the matter of worker resistance, with employees complaining that they do not want to work in a "Big Brother" environment of constant surveillance.[22]

Schools, however, are not hospitals, and most adults would happily keep their kids under constant surveillance. Active RFID systems are gradually being deployed in schools. Children are issued chip-enhanced ID badges that signal a badge reader mounted on the schoolhouse door or the entrance to the school bus. Simply by passing through the portal, the student notifies school officials that he or she is in the right place at the right time. The system can also fire off an e-mail to the parents to confirm that their child made it to school. "We don't want it to be Big Brother," said Kayode Aladesuyi, the chief executive officer of StudentConnect, of Marietta, Georgia, a vendor of RFID systems for schools. "We want it to be information that parents are using for the enhancement of the kids' education and for their safety." For instance, if a child gets off the school bus but has not entered the building within a few minutes, school officials receive an immediate alert. Aladesuyi also cited cases in which small children were left behind on school buses because drivers had lost track of how many had boarded. An RFID-equipped bus would ensure this never happened, according to Aladesuyi. By combining RFID with a GPS system to track the bus itself, administrators and parents could know the child's exact location during the commute to and from the school.[23]

At John Jay High School in San Antonio, Texas, RFID tracking was embraced to protect the students and the district's budget. State funding is based on daily attendance, as measured by an old-fashioned roll call. If a child is late for the head count, the school gets less money, but a school can increase its income by thousands of dollars per year by making sure every child is counted. With an RFID system, that becomes child's play; as soon as a kid walks through the door, he is logged in.

Drawn by the lure of extra funding, two of the district's schools adopted an RFID system in the autumn of 2012. Administrators expected no objections from students or parents. More important, they did not count on the Hernandez family, a devoutly religious brood of a fundamentalist bent. Brought up on a literalistic reading of the biblical book of Revelation, sophomore Andrea Hernandez believed that the RFID tag was "the mark of the beast" described in chapter 13 of that apocalyptic work. This beast, one of two blasphemous enemies of God described in the chapter, puts a mark on the hand or forehead of all people. Those who refuse the mark cannot buy or sell anything, making it an offer that is extremely difficult to refuse. Accepting the mark of the beast amounts to pledging allegiance to God's enemies, while refusing to accept it makes one an economic and social exile. Believers are warned in chapter 14 that any who accept the mark "shall drink of the wine of the wrath of God, which is poured out without mixture into the cup of his indignation; and he shall be tormented with fire and brimstone in the presence of the holy angels, and in the presence of the Lamb" (Rev. 14:10–11).

It seems like a greater weight of cosmic significance than a plastic ID badge can bear. Yet the Hernandez family was convinced that an order to wear an RFID tag amounted to an effort to force the mark of the beast on Andrea. In August 2012, Andrea's father, Steven, wrote to the school district, saying that "it is our Hell Fire Belief that if we compromise our faith and religious freedom to allow you to track my daughter while she is at school it will condemn us to hell."[24]

Perhaps because fervent religious belief is not rare in Texas, the school board offered a substantial compromise. They agreed to issue Andrea an ID badge that would not contain an RFID chip, rendering her movements untraceable. It was not good enough for the Hernandez family, who said that even without the chip inside, wearing the badge would imply that they approved of a school policy that they still considered sinful. Andrea was warned that she would be expelled unless she complied. Her family went to court and in November 2012 won a temporary restraining order against the school district. In January 2013 a federal judge ruled that Hernandez must wear the badge or seek another school. Hernandez transferred to a different San Antonio high school, one free of RFID badges.

But Hernandez has since returned to John Jay High. In July 2013 the school district announced that it would stop using the badges, having decided that the RFID system did not measure up to their expectations. It is not exactly a great victory for privacy, though. The school has instead placed video cameras throughout the school—two hundred of them.[25]

Few students or parents share the Hernandez family's fervent religious objections to RFID, but they still mustered plenty of support from civil liberties organizations, including the Electronic Frontier Foundation, the Electronic Privacy Information Center, and the American Civil Liberties Union. These groups worry that constant tracking is demeaning to students, that it may dissuade them from seeking advice from school counselors for fear that their visits will be monitored, that unauthorized persons armed with RFID readers could monitor students even when they leave the campus.[26]

Perhaps the most troubling aspect is the way that constant tracking of students conditions young people to expect similarly intrusive surveillance as adults. "Regardless of the intended purpose, the message for students is to cooperate with this extended control and get used to it," write two critics. "In academic language, we would say that students are 'normalized' to this surveillance—it becomes commonplace, unquestioned and unremarkable."[27]

Indeed, the pervasive use of multiple location-aware technologies makes it difficult to imagine a world in which your whereabouts are a secret to everybody but yourself. Even dedicated and capable data privacy mavens such as Bruce Schneier, author of multiple books on computer security, have thrown in the towel. "The Internet is a surveillance state," laments Schneier. "Whether we admit it to ourselves or not, and whether we like it or not, we're being tracked all the time." Between them police and marketing companies are compiling every important fact about us, including where we happen to be, and even where we probably will be in two hours, to do with as they see fit. "Welcome to a world where all of this, and everything else that you do or is done on a computer, is saved, correlated, studied, passed around from company to company without your knowledge or consent," said Schneier, "and where the government accesses it at will without a warrant."

However, it is too early to despair. Locational privacy, the right to keep one's whereabouts secret, has found champions across the political spectrum. In 2013 Republicans and Democrats in both houses of the US Congress introduced bills that would force police to get a probable-cause warrant to electronically track a citizen's location and movements. The new laws cover real-time location tracking, but they would also require a warrant for access to the location data in someone's old cell phone records.[28]

Law enforcement agencies warn that the tougher standards would cripple them. In testimony to a congressional hearing in April 2013, Peter A. Modafferi, chairman of the police investigative operations committee of the International Association of Chiefs of Police, cited a case where narcotics detectives in Rockland County, New York, received a tip about a person who was running a drug lab. The information was not enough to establish probable cause. However, by obtaining the suspect's location data from his cell phone provider, the detectives were able to figure out the address of the lab and stake out the place. The surveillance produced enough evidence of drug activity to get them a search warrant. Yet Modafferi said that warrantless access to the suspect's location data played a vital role in making the case. "Requiring probable cause to get basic, limited information about a person's historical location would make it significantly more difficult to solve crimes and seek justice," said Modafferi. Possibly. Many lawmakers dread the alternative: a world in which police agencies can obtain at will detailed information on the location and movements of every citizen.

There is also action at the state level. In 2013 legislatures in Texas and Maryland, for instance, took up bills that would require law enforcement officers to obtain wiretap-style probable-cause warrants before they could get location data on citizens. However, the cause of personal privacy would be ill-served by fifty different legal standards for locational privacy. The United States needs a unified federal standard. It ought to cover both real-time data obtained by live tracking of citizens and access to historical location data on file with cell phone carriers and technology companies that track user location, like Google and Apple.

We also need better oversight of the location data collected by the private sector. Senator Al Franken, a Minnesota Democrat, drafted the Location Privacy Protection Act. This law would require that any software on a mobile communications device must explicitly get the user's permission before recording his location. Thus, an app that wants to track you must ask you first. Additionally, the law would require software makers to get a user's permission to share his location data with any other company.[29]

Yet it may not be enough to rely on law alone. We would be better off with location-based systems that have privacy features built into the software. Consider the EZ Pass, which helps millions glide quickly through tollbooths, but also records the fact that Jane Doe was headed west on the Massachusetts Turnpike last Tuesday evening. There is no reason the system has to know the identity of the users. Andrew Blumberg, assistant professor of mathematics at the University of Texas at Austin, suggests redesigning the system as a stored-value service. Instead of sending a bill to the home of an EZ Pass user, the pass would be linked to an account containing a certain amount of money. The tollway authority would immediately collect its cash from each user. At regular intervals the user would top up the account by adding more money. This could be done at a bank or retail store, but Blumberg prefers the idea of using an anonymous "electronic cash" service like the controversial and little-used Bitcoin system. If an EZ Pass service accepted Bitcoin, a user could add value to the account from any computer, in a manner that would leave behind no trace of his or her identity.[30]

Aside from the fact that hardly anyone uses Bitcoin, it is difficult to see how this EZ Pass revision would pass muster with police and Homeland Security agencies. They have grown accustomed to using EZ Pass to track down lawbreakers and are not likely to favor a less useful redesign of the system. And law enforcement officials are not the only ones who have grown accustomed to the existing system; so have drivers. Most have come to tolerate the attendant loss of privacy in exchange for efficiency and convenience.

Although EZ Pass is not likely to see a useful redesign, other ways to protect driver privacy are still open to us. Back in 2005 Blumberg and several colleagues argued that there are good reasons to put radio transponders on all cars, to help governments manage traffic and track down dangerous drivers. Such transponders could also be used for automatic toll payments, like EZ Pass. But they would also be a privacy nightmare, effectively allowing governments to track every trip made by every motorist. Yet it is possible to build such a system that would track each vehicle anonymously, so that the police and highway departments would get useful information without knowing the identities of individual drivers—except when they needed to.

The scheme would allow each car's transponder to transmit a different identification code every few seconds, from a library containing many millions of code numbers. Under normal circumstances, it would be impossible to know the individual driver represented by a particular code. Still, the authorities could monitor traffic by tracking the location and speed of the moving vehicle, in much the same way that Google and Apple already do.

Suppose a driver runs a red light. A data capture device attached to the light would merely record the identification code being transmitted by the car at that moment. Police could look up the identity of the person associated with that code at that time and send a traffic citation in the mail. However, seconds after the offense, the driver's code will have changed. The police would have no way of tracking her movements either before or after she ran the red light. Such a system could achieve the government's legitimate public safety goals, while protecting the public from the threat of incessant surveillance.[31]

Other researchers, like IBM's Jeff Jonas, are working on even more expansive solutions that could let people share their location data and obscure it at the same time. The information could be used to provide geographical guidance to consumers, but the business providing the information would not be given accurate information about the customer's location. For example, if I ask Google Maps for directions to a hotel, Google needs a starting point to provide guidance. These days Google starts with my latitude and longitude, provided by GPS or Wi-Fi tracking. In a more privacy-sensitive system, by contrast, this information would be replaced by a computer-generated code that would provide my approximate location. "They would know I'm near a sandwich shop, but they wouldn't actually have the latitude and longitude," said Jonas. The system would encrypt the user's precise location data, making it impossible to accurately trace any unwilling user.

Jonas has not perfected his system. Even if he does, it is hardly a panacea. Cellular carriers must know their customers' locations to route their calls; police agencies will always be able to get hold of that information. But by combining anonymizing technology with tougher legal limits on access to location data, each of us might be able to preserve for ourselves a cocoon of locational privacy.

It is a solution that just might work, but only at the cost of constant vigilance. And that is an ironic reversal of fortune. For most of human history, it has been relatively easy to get lost; keeping track of people and their movements was the great challenge, and solving it has been the work of millennia. Over the past century we have finally reached the goal, but our success has come at a price. Maps of the world used to feature mountains, oceans, and skyscrapers. With today's maps the primary geographic feature is us—the exact locations of billions of humans, their every move recorded and viewed by far too many eyes. Smarter technologies and better laws may protect us from the worst excesses of police spies and corporate marketers. But never again will we be shielded by ignorance. They know where to find us, and they always will.

Acknowledgments

THE KNOW-IT-ALL MOTORIST ON A CROSS-COUNTRY TRIP MIGHT TRAVEL A hundred miles out of his way rather than stop to ask for directions. But such arrogance is a luxury no author can afford. Forget the hundreds of solitary hours at a keyboard; nobody really writes alone, especially not when tackling such a complex topic as navigation. So I'm grateful to the many people who've guided me on my journey.

Bradford Parkinson, who led the development of the Global Positioning System, was generous with his time. So were GPS engineer Tom Logsdon, Boston patent attorney and GPS innovator Robert Tendler, and Stephen Poizner, whose SnapTrack technology helped turn billions of phones into personal navigators. Stéphane Dubois, chief of communications at the International Civil Aviation Organization, provided essential documents about the navigational blunders that led to the tragic 1983 shootdown of Korean Air Lines Flight 007. Ted Morgan, co-founder of Skyhook, helped me understand how Wi-Fi routers became homing beacons, while former Apple Inc. executive Robert Borchers provided crucial information on Apple's decision to embrace Skyhook's remarkable innovation. Steve Coast, the founder of the OpenStreetMap project; Joshua Stanton of the One Free Korea blog; Curtis Melvin, editor of North Korean Economy Watch; and Erik Hersman, the co-founder of Ushahidi, taught me much about do-it-yourself cartography. John Hanke, founder of Keyhole Inc., and Walter Scott, founder and chief technical officer of DigitalGlobe, showed me how images shot from space have given us extraordinary new ways to map the earth.

My research into the commercial uses of location technology benefited from the generous assistance of Joe Francica, editor-in-chief of *Directions Magazine;* David Petersen, chief executive officer of Sense Networks; Placecast chief executive Alistair Goodman; and Asif Khan, president of the Location Based Marketing Association. Privacy advocate Katherine Albrecht; cryptographer and computer security analyst Bruce Schneier; Kayode Aladesuyi, the chief executive officer of StudentConnect, Inc.; and IBM Corp. Fellow Jeff Jonas were most helpful as I researched the privacy issues raised by our mastery of location tracking.

I'm grateful to my agent, Michelle Tessler, who shared my fervent interest in navigation and saw the promise in my first crude efforts to tell this story. I'm also indebted to Tisse Takagi, my patient and insightful editor at Basic Books.

The Men's Fellowship at Peoples Baptist Church offered the moral support that every writer craves. And if it weren't for the loyal encouragement of my wife Dimonika, and my daughters Ariana and Octavia, you'd be reading something else right now, and perhaps enjoying it less. God bless the lot of them.

Notes

Chapter 1

1. Nathaniel Philbrick, *Mayflower: A Story of Courage, Community, and War* (New York: Penguin Books, 2006), 33–46.

2. Fiona Govan, "World's Oldest Map: Spanish Cave Has Landscape from 14,000 Years Ago," *Daily Telegraph*, August 6, 2009.

3. David Lewis, *We, the Navigators: The Ancient Art of Landfinding in the Pacific* (Honolulu: University of Hawaii Press, 1994), 124–133, www.amazon.com/David-Lewis/e/B001HPYLXO/ref=ntt_athr_dp_pel_1.

4. Leo Bagrow, *History of Cartography* (Cambridge, MA: Harvard University Press, 1964), 27–28.

5. Amir D. Aczel, *The Riddle of the Compass* (New York: Harcourt, 2001), 16–17.

6. Eva Germaine Rimington Taylor, *The Haven-Finding Art: A History of Navigation from Odysseus to Captain Cook* (London: Hollis & Carter, 1956), 8–14.

7. Lloyd Motz and Jefferson Hane Weaver, *The Story of Astronomy* (New York: Plenum Press, 1995), 45.

8. Jerry Brotton, *Trading Territories: Mapping the Early Modern World* (Ithaca, NY: Cornell University Press, 1997), 54.

9. Lloyd A. Brown, *The Story of Maps* (Boston: Little, Brown, 1950), 59–80.

233

10. John Noble Wilford, *The Mapmakers* (New York: Alfred A. Knopf, 2000), 32–36.

11. A. R. T. Jonkers, *Earth's Magnetism in the Age of Sail* (Baltimore: Johns Hopkins University Press, 2003), 66.

12. Wilford, *The Mapmakers*, 64.

13. Ibid., 87–104.

14. Andrew D. Lambert, *The Gates of Hell: Sir John Franklin's Tragic Quest for the North West Passage* (New Haven, CT: Yale University Press, 2009), 74.

15. Frank Nothen Magill, *Magill's Survey of Science: Earth Science Series* (Pasadena, CA: Salem Press, 1990), 2:541.

16. Dava Sobel, *Longitude: The True Story of a Lone Genius Who Solved the Greatest Scientific Problem of His Time* (New York: Penguin Books, 1995), 24–27.

17. Wilford, *The Mapmakers*, 132–151.

Chapter 2

1. Abigail Foerstner, *James Van Allen: The First Eight Billion Miles* (Iowa City: University of Iowa Press, 2007), 50.

2. US Patent 716134, issued December 16, 1902.

3. Captain Linwood S. Howeth, USN (Retired), *History of Communications-Electronics in the United States Navy* (Washington, DC: US Government Printing Office, 1963).

4. U. A. Bakshi, A. V. Bakshi, and K. A. Bakshi, *Antenna and Wave Propagation* (Pune, India: Technical Publications, 2009), 3–20.

5. "Over-Ocean Phone Halted by the War," *New York Times*, October 23, 1914, 9.

6. Monte D. Wright, *Most Probable Position: A History of Aerial Navigation to 1941* (Lawrence: University Press of Kansas, 1972), 22.

7. Ibid., 89.

8. Douglas H. Robinson, *The Zeppelin in Combat* (London: G. T. Foulis, 1966), 72.

9. "How the Zeppelin Raiders Are Guided by Radio Signals," *Popular Science Monthly*, April 1918, 632–634, http://earlyradiohistory.us/1918zep.htm.

10. Russell Burns, *Communications: An International History of the Formative Years* (London: Institution of Electrical Engineers, 2004), 421.

11. Jeffery T. Richelson, *A Century of Spies: Intelligence in the Twentieth Century* (New York: Oxford University Press, 1997), 40.

12. David Kahn, *The Codebreakers: The Story of Secret Writing* (New York: Simon & Schuster, 1996), 273.

13. F. Robert van der Linden, *Airlines and Airmail* (Lexington: University Press of Kentucky, 2002), 5–6.

14. Carroll V. Glines, *Airmail: How It All Began* (Blue Ridge Summit, PA: Tab Books, 1990), 34.

15. Erik M. Conway, *Blind Landings: Low-Visibility Operations in American Aviation, 1918–1958* (Baltimore: Johns Hopkins University Press, 2006), 187.

16. Ibid., 18.

17. Barry Rosenberg and Catherine Macaulay, *Mavericks of the Sky: The First Daring Pilots of the US Airmail* (New York: William Morrow, 2006), 245–246.

18. Van der Linden, *Airlines and Airmail*, 10.

19. Wright, *Most Probable Position*, 120.

20. Ibid., 126.

21. Nikola Tesla, "The Problem of Increasing Human Energy," *Century Illustrated*, June 1900. See also Margaret Cheney, *Tesla: Man Out of Time* (New York: Touchstone Books, 2001), 259–260.

22. "Tesla's Views on Electricity and the War," *Electrical Experimenter*, August 1917.

23. John Waller, *Leaps in the Dark: The Forging of Scientific Reputations* (New York: Oxford University Press, 2004), 224–225.

24. Robert Buderi, *The Invention That Changed the World* (New York: Simon and Schuster, 1996), 61–64.

25. Ibid., 27–28.

26. Ibid., 56.

27. Kathleen Broome Williams, *Secret Weapon: US High-Frequency Direction Finding in the Battle of the Atlantic* (Annapolis, MD: Naval Institute Press, 1996), 65.

28. Louis Brown et al., *The Department of Terrestrial Magnetism*, vol. 2 of *Centennial History of the Carnegie Institution of Washington* (Cambridge: Cambridge University Press, 2004), 110.

29. Milton Friedman and Rose D. Friedman, *Two Lucky People: Memoirs* (Chicago: University of Chicago Press, 1999), 136–137.

30. Foerstner, *James Van Allen*, 54–55.

31. Buderi, *Invention That Changed the World*, 228.

32. Charles B. MacDonald, *Victory in Europe, 1945: The Last Offensive of World War II* (Mineola, NY: Courier Dover, 2007), 13.

33. Dr. Merle Tuve, interview by Albert Christman, Terrestrial Magnetism Laboratory, Washington, DC, May 6, 1967, www.aip.org/history/ohilist /3894.html.

Chapter 3

1. YouTube features some nice videos of Foucault pendulums in action. For instance, www.youtube.com/watch?v=b14l3-A8iUQ.

2. "South Pole Foucault Pendulum," www.phys-astro.sonoma.edu/graduates /baker/southpolefoucault.html.

3. Jörg F. Wagner and Andor Trierenberg, "The Machine of Bohnenberger: Bicentennial of the Gyro with Cardanic Suspension," *Proceedings in Applied Mathematics and Mechanics* 10, no. 1 (2010): 659–660.

4. Amir D. Aczel, *Pendulum: Leon Foucault and the Triumph of Science* (New York: Atria Books, 2003), 165–171.

5. Thomas Parke Hughes, *Elmer Sperry: Inventor and Engineer* (Baltimore: Johns Hopkins University Press, 1971), 130–131.

6. Donald MacKenzie, *Inventing Accuracy: A Historical Sociology of Nuclear Missile Guidance* (Cambridge: MIT Press, 1993), 34.

7. Richard Knowles Morris, *John P. Holland, 1841–1914: Inventor of the Modern Submarine* (Columbia: University of South Carolina Press, 1998), 58.

8. US Patent 907907, issued December 29, 1908.

9. Hughes, *Elmer Sperry*, 122.

10. Matthew Trainer, "Albert Einstein's Expert Opinions on the Sperry vs. Anschütz Gyrocompass Patent Dispute," *World Patent Information* 30, no. 4 (2008): 320–325.

11. Hughes, *Elmer Sperry*, 262.

12. Kenneth P. Werrell, *The Evolution of the Cruise Missile* (Maxwell Air Force Base, AL: Air University Press, 1985). See also Lee Pearson, "Developing the Flying Bomb," in *Naval Aviation in World War I*, edited by Adrian O. Van Wyen (Washington, DC: Chief of Naval Operations, 1969).

13. Fred H. Previc and William R. Ercoline, *Spatial Disorientation in Aviation* (Reston, VA: American Institute of Aeronautics and Astronautics, 2004), 7–8.

14. Carroll V. Glines, *Jimmy Doolittle: Daredevil Aviator and Scientist* (New York: Macmillan, 1972), 75–91.

15. Joseph John Murphy, "Instinct: A Mechanical Analogy," *Nature*, April 24, 1873, 483.

16. Donald J. Clausing, *The Aviator's Guide to Navigation* (New York: McGraw-Hill Professional, 2006), 167–171.

17. MacKenzie, *Inventing Accuracy*, 46–47.

18. Martin J. Collins and Martin Collins, *After Sputnik: 50 Years of the Space Age* (New York: HarperCollins, 2007), 22.

19. *Effects of Nuclear Earth-Penetrator and Other Weapons* (Washington, DC: National Research Council, 2005), 5.

20. MacKenzie, *Inventing Accuracy*, 67.

21. Ibid., 76–78.

22. Lawrence Freedman, *The Evolution of Nuclear Strategy* (New York: Palgrave Macmillan, 1989), 24.

23. Ibid., 335.

24. Ibid., 378–406.

25. Michael S. Nolan, *Fundamentals of Air Traffic Control* (Clifton Park, NY: Cengage Learning, 2010), 75–77. See also Art Zuckerman, "Doppler Radar Charts the Airlanes," *Popular Electronics*, May 1959, 41.

26. "Inertial for the 747," *Flight International*, September 4, 1969.

27. MacKenzie, *Inventing Accuracy*, 180–182.

Chapter 4

1. Matthew Brzezinski, *Red Moon Rising: Sputnik and the Hidden Rivalries That Ignited the Space Age* (New York: Macmillan, 2007), 161–187.

2. William H. Guier and George C. Weiffenbach, "Genesis of Satellite Navigation," *Johns Hopkins APL Technical Digest* 19, no. 1 (1998): 14–17.

3. Helen Gavaghan, *Something New Under the Sun* (New York: Springer-Verlag, 1998), 87.

4. Mike Gruntman, *Blazing the Trail: The Early History of Spacecraft and Rocketry* (Reston, VA: American Institute of Aeronautics and Astronautics, 2004), 196.

5. Graham Spinardi, *From Polaris to Trident: The Development of US Fleet Ballistic Missile Technology* (Cambridge: Cambridge University Press, 1994), 29–34.

6. Donald A. MacKenzie, *Inventing Accuracy: A Historical Sociology of Nuclear Missile Guidance* (Cambridge: MIT Press, 1993), 142–148.

7. Edward Everett Hale, "The Brick Moon," www.readbookonline.net/read OnLine/2108/.

8. Gavaghan, *Something New Under the Sun*, 89.

9. Ibid., 109.

10. Teddy Seymour, "No-Frills Navigation," www.bluemoment.com/seymour .html.

11. Bradford W. Parkinson et al., "A History of Satellite Navigation," *Navigation* 42, no. 1 (1995).

12. Jacob Neufeld, George M. Watson Jr., and David Chenoweth, eds., *Technology and the Air Force: A Retrospective Assessment* (Washington, DC: Air Force Historical Studies Office, Bolling Air Force Base, 1997), 247–248.

Chapter 5

1. "Cell Phone Saves Woman Trapped in Blizzard," Associated Press, January 12, 1997.

2. "Kingwood Deaths Spur Call for Better 911 Service," *Houston Chronicle*, April 2, 2000.

3. "'Smart Cars' Will Make Crash Victims Easier to Find," *USA Today*, March 6, 2001.

4. Ivan Getting, *All in a Lifetime: Science in the Defense of Democracy* (New York: Vantage Press, 1989), 578–580. See also Jacob Neufeld, George M. Watson Jr., and David Chenoweth, *Technology and the Air Force: A Retrospective Assessment, Air Force History and Museums Program* (Washington, DC: United States Air Force, 1997), 248.

5. Mike Gruntman, *Blazing the Trail: The Early History of Spacecraft and Rocketry* (Reston, VA: American Institute of Aeronautics and Astronautics, 2004), 350.

6. Ann Darrin and Beth L. O'Leary, eds., *Handbook of Space Engineering, Archaeology, and Heritage* (Boca Raton, FL: CRC Press, 2009), 244.

7. "Leap Seconds," US Naval Observatory, http://tycho.usno.navy.mil /leapsec.html.

8. Michael A. Lombardi, Thomas P. Heavner, and Steven R. Jefferts, "NIST Primary Frequency Standards and the Realization of the SI Second," *Measure: The Journal of Measurement Science* (December 2007): 74–89.

9. James Jespersen and Jane Fitz-Randolph, *From Sundials to Atomic Clocks: Understanding Time and Frequency* (New York: Dover, 1982), 39–46. For a description of the most accurate cesium clock, see the National Institute of Standards and Technology website, www.nist.gov/pml/div688/grp50/primary -frequency-standards.cfm.

10. Donald J. Clausing, *The Aviator's Guide to Navigation* (New York: McGraw-Hill Professional, 2006), 155–165.

11. Neufeld, Watson, and Chenoweth, *Technology and the Air Force*, 250. See also Naval Research Laboratory, "2008 Review," April 2009, 12, www.nrl .navy.mil/content_images/2008_NRL_Review.revised.pdf.

12. Stephen T. Powers and Bradford Parkinson, "The Origins of GPS, Part I," *GPS World,* May 1, 2010.

13. Getting, *All in a Lifetime,* 588.

14. Astronautix, "SECOR," www.astronautix.com/craft/secor.htm.

15. Dana J. Johnson, "Overcoming Challenges to Transformational Space Programs: The Global Positioning System (GPS)," Analysis Center, Northrup Grumman, 2006, 8, www.analysiscenter.northropgrumman.com/files/NGACP _1006d.pdf.

16. Neil Ashby, "General Relativity in the Global Positioning System," www .leapsecond.com/history/Ashby-Relativity.htm. See also Clifford M. Will, "Einstein's Relativity and Everyday Life," www.physicscentral.com/explore/writers /will.cfm.

17. Jim Schefter, "Geostar," *Popular Science,* June 1984, 76. See also www .rdss.com/.

18. 8 Scott Pace et al., *The Global Positioning System: Assessing National Policies* (Santa Monica, CA: RAND, 1995), 264.

19. "Crew Uses GPS to Cross Ocean," *Aviation Week and Space Technology,* June 6, 1983.

20. Powers and Parkinson, "Origins of GPS."

21. Asaf Degani, *Taming HAL: Designing Interfaces Beyond 2001* (New York: Palgrave Macmillan, 2004), 49–65. See also *Destruction of Korean Air Lines Boeing 747 on 31 August 1983* (Montreal: International Civil Aviation Organization, June 1993), 42.

22. *Federal Register,* 46 F.R. 20724, April 7, 1981.

23. Tom Logsdon, interview with the author, August 23, 2012.

24. United States National Imagery and Mapping Agency, *The American Practical Navigator* (Arcata, CA: Paradise Cay, 2002), 167. See also Pace et al., *Global Positioning System,* 86.

25. Joel McNamara, *GPS for Dummies* (New York: John Wiley and Sons, 2008), 58.

26. Shirish Date, "Mixed Signals on Use of Navigation Satellites," *Orlando Sentinel,* March 18, 1993.

27. Federal Aviation Administration, "GNSS Frequently Asked Questions: WAAS," www.faa.gov/about/office_org/headquarters_offices/ato/service _units/techops/navservices/gnss/faq/waas/index.cfm.

28. Vincent Kiernan, "Guidance from Above in the Gulf War," *Science,* March 1, 1991, 1012–1014.

29. John Holusha, "Lost in Yonkers? An Olds Option Could Be a Guide," *New York Times,* November 27, 1994.

30. Dorinda G. Dallmeyer and Kosta Tsipis, *Heaven and Earth: Civilian Uses of Near-Earth Space* (The Hague: Kluwer Law International, 1997), 207–208.

31. Steven Kettmen, "Europe Gives Go-Ahead to Galileo," *Wired*, March 18, 2002, www.wired.com/news/politics/0,1283,51130,00.html.

32. Roftiel Constantine, *GPS and Galileo: Friendly Foes?* (Maxwell Air Force Base, AL: Air University Press, 2007), 9.

33. US Department of Defense, "DoD Permanently Discontinues Procurement of Global Positioning System Selective Availability," September 18, 2007, www.defense.gov/Releases/Release.aspx?ReleaseID=11335.

34. Michael A. Lombardi and Wayne Hanson, "The GOES Time Code Service, 1974–2004: A Retrospective," *Journal of Research of the National Institute* (March–April 2005).

35. "PCS Will Add to Problem; Wireless Industry Grapples with Concerns About 911 Incompatibility," *Communications Daily*, March 28, 1994.

36. "GPS and U-TDOA: Frequently Asked Questions," 2009, www.true position.com/frequently-asked-questions-about-u-tdoa/DownloadSecured.

37. Hiawatha Bray, "Museum Sees Him as GPS Pioneer; He Hopes Patent Suit Judge Agrees," *Boston Globe*, April 14, 2007.

38. Steve Poizner, interview with the author, October 6, 2012.

39. Nam D. Pham, "The Economic Benefits of Commercial GPS Use in the U.S. and the Costs of Potential Disruption," June 2011, 2, www.saveourgps.org /pdf/GPS-Report-June-22–2011.pdf.

40. Rick W. Sturdevant, "Tracing Connections—Vanguard to NAVSPASUR to GPS: An Interview with Roger Lee Easton, Sr.," *High Frontier: The Journal for Space and Missile Professionals* (May 2008): 54.

41. Bradford Parkinson, e-mail message, October 30, 2013.

Chapter 6

1. Wolter Lemstra, Vic Hayes, and John Groenewegen, *The Innovation Journey of Wi-Fi: The Road to Global Success* (Cambridge: Cambridge University Press, 2010), 21–52.

2. August E. Grant and Jennifer H. Meadows, *Communication Technology Update and Fundamentals* (Boca Raton, FL: CRC Press, 2013), 83.

3. Richard Rhodes, *Hedy's Folly* (New York: Random House, 2012).

4. Harold Abelson, Ken Ledeen, and Harry R. Lewis, *Blown to Bits: Your Life, Liberty, and Happiness After the Digital Explosion* (Upper Saddle River, NJ: Addison Wesley Professional, 2008), 286–288.

5. Lemstra, Hayes, and Groenewegen, *Innovation Journey of Wi-Fi*, 121–125.

6. www.businesswire.com/news/home/20121011006017/en/Wi-Fi-Enabled -Device-Shipments-Exceed-1.5-Billion.

7. Ted Morgan, interview with the author, July 9, 2011.

8. Bob Borchers, interview with the author, July 24, 2011.

9. Peter Farago, "Day 74 Sales: Apple iPhone vs. Google Nexus One vs. Motorola Droid," Flurry Blog, March 16, 2010, http://blog.flurry.com /bid/31410/Day-74-Sales-Apple-iPhone-vs-Google-Nexus-One-vs -Motorola-Droid.

10. Peter Fleischer, "Data Collected by Google Cars," Google Europe Blog, April 27, 2010, http://googlepolicyeurope.blogspot.com/2010/04/data -collected-by-google-cars.html.

11. Alan Eustace, "WiFi Data Collection: An Update," Google Official Blog, May 14, 2010, *http://googleblog.blogspot.com/2010/05/wifi-data-collection -update.html*.

12. *Skyhook Wireless, Inc. v. Google Inc.*, 1:10-cv-11571-RWZ, US District Court, District of Massachusetts; *Skyhook Wireless, Inc. v. Google Inc.*, 2010-03652-BLS1, Suffolk Superior Court, Commonwealth of Massachusetts.

Chapter 7

1. Curtis Peebles, *Shadow Flights: America's Secret Air War Against the Soviet Union* (Novato, CA: Presidio Press, 2000), 261–271.

2. Robert D. Mulcahy Jr., ed., *Corona Star Catchers: The Air Force Aerial Recovery Aircrews of the 6593d Test Squadron (Special), 1958–1972* (Chantilly, VA: Center for the Study of National Reconnaissance, National Reconnaissance Office, June 2012), 1–14.

3. Grover Heiman, *Aerial Photography: The Story of Aerial Mapping and Reconnaissance* (New York: Macmillan, 1972), 8.

4. John Noble Wilford, *The Mapmakers* (New York: Alfred A. Knopf, 2000), 271.

5. William E. Burrows, *Deep Black: Space Espionage and National Security* (New York: Random House, 1986), 28–30.

6. Heiman, *Aerial Photography*, 22–24, 31.

7. Richard P. Hallion, *Taking Flight: Inventing the Aerial Age, from Antiquity Through the First World War* (New York: Oxford University Press, 2003), 310–312.

8. Ibid., 355.

9. Dino A. Brugioni, *Eyes in the Sky: Eisenhower, the CIA, and Cold War Aerial Espionage* (Annapolis, MD: Naval Institute Press, 2010), 3.

10. Heiman, *Aerial Photography*, 58–59. See also "Biography of Brig. Gen. George William Goddard, US Air Force," www.af.mil/information/bios/bio.asp?bioID=5563.

11. Brugioni, *Eyes in the Sky*, 6.

12. John F. Kreis et al., *Piercing the Fog: Intelligence and Army Air Forces Operations in World War II* (Washington, DC: Air Force History and Museums Program, Bolling Air Force Base, 1996), 80–82.

13. Brugioni, *Eyes in the Sky*, 22.

14. Andrew Curry, "German Firm Uses Aerial Photos to Find Bombs," *Der Spiegel*, April 9, 2012.

15. Birger Stichelbaut et al., eds., *Images of Conflict: Military Aerial Photography and Archaeology* (Newcastle upon Tyne, England: Cambridge Scholars, 2009), 6–10.

16. Brugioni, *Eyes in the Sky*, 33.

17. Burrows, *Deep Black*, 59.

18. Peebles, *Shadow Flights*, 8–10.

19. Joseph A. Angelo, *Nuclear Technology* (Westport, CT: Greenwood, 2004), 58–60.

20. John C. Baker, Kevin M. O'Connell, and Ray A. Williamson, eds., *Commercial Observation Satellites: At the Leading Edge of Global Transparency* (Santa Monica, CA: RAND, 2001), 18–19.

21. Brugioni, *Eyes in the Sky*, 141–145.

22. Peebles, *Shadow Flights*, 89.

23. Ibid., 149.

24. Stephen I. Schwartz, *Atomic Audit: The Costs and Consequences of U.S. Nuclear Weapons Since 1940* (Washington, DC: Brookings Institution Press, 1998), 233.

25. *Preliminary Design of an Experimental World-Circling Spaceship* (Santa Monica, CA: RAND, 1946), 13–14.

26. *Utility of a Satellite Vehicle for Reconnaissance* (Santa Monica, CA: RAND, 1951).

27. Curtis Peebles, *High Frontier: The US Air Force and the Military Space Program* (Washington, DC: Air Force History and Museums Program, 1997), 6–7.

28. Ibid., 9.

29. Philip Taubman, *Secret Empire: Eisenhower, the CIA, and the Hidden Story of America's Space Espionage* (New York: Simon and Schuster, 2003), 237–238.

30. Baker, O'Connell, and Williamson, *Commercial Observation Satellites*, 30.

31. Jeffrey T. Richelson, *America's Secret Eyes in Space: The US Keyhole Spy Satellite Program* (New York: Harper and Row, 1990), 97–98.

32. Mark Monmonier, *Spying with Maps: Surveillance Technologies and the Future of Privacy* (Chicago: University of Chicago Press, 2004), 30.

33. Ben Iannotta, "Spy-Sat Rescue," *Defense News* (June 2, 2009). See also Ronald O'Rourke, "Navy CVN-21 Aircraft Carrier Program: Background and Issues for Congress," Navy Department Library, www.history.navy.mil /library/online/navycvn21.htm.

34. Pamela Etter Mack, *Viewing the Earth: The Social Construction of the Landsat Satellite System* (Cambridge: MIT Press, 1990), 60–61.

35. Baker, O'Connell, and Williamson, *Commercial Observation Satellites*, 40–47.

36. James N. Rosenau and J. P. Singh, *Information Technologies and Global Politics: The Changing Scope of Power and Governance* (Albany: State University of New York Press, 2002), 76–77.

37. Walter Scott, interview with the author, November 26, 2012.

Chapter 8

1. For an excellent historical summary, see Michael Friendly and D. J. Denis, "Milestones in the History of Thematic Cartography, Statistical Graphics, and Data Visualization," www.datavis.ca/milestones.

2. Panel on Research on Future Census Methods, National Research Council, *Planning the 2010 Census: Second Interim Report* (Washington, DC: National Academies Press, 2003), 46–47.

3. The Google Maps API site, www.morethanamap.com.

4. NielsenWire, May 25, 2012, http://blog.nielsen.com/nielsenwire/online _mobile/april-2012-top-u-s-online-brands-and-travel-websites.

5. Jerry Brotton, *A History of the World in Twelve Maps* (London: Allen Lane, 2012), 417–418.

6. Kevin Maney, "Tiny Tech Company Awes Viewers," *USA Today*, March 21, 2003.

7. John Timmer, "New Satellite to Give Google Maps Unprecedented Resolution," *Ars Technica*, http://arstechnica.com/business/2008/09/new-satellite -to-give-google-maps-unprecedented-resolution/.

8. Steven Levy, "The Earth Is Ready for Its Close-Up," *Newsweek*, June 6, 2005, 13.

9. UNOSAT Humanitarian Rapid Mapping Service, "Overview 2011," http://unosat.web.cern.ch/unosat/unitar/Overview2011UNOSATRapid Mapping_final2.pdf.

10. Danny Bradbury, "Taking Your Network to Extremes," *Computer Weekly,* March 27, 2007.

11. Patrick Meier, "The Past and Future of Crisis Mapping," iRevolution, October 18, 2008, http://irevolution.net/2008/10/18/future-of-crisis-mapping/.

12. Clay Shirky, *Cognitive Surplus: Creativity and Generosity in a Connected Age* (New York: Penguin Press, 2010), 15–17.

13. Andrew Zolli and Ann Marie Healy, *Resilience: Why Things Bounce Back* (New York: Simon and Schuster, 2012), 172–190.

14. Jessica Heinzelman and Carol Waters, "Crowdsourcing Crisis Information in Disaster-Affected Haiti," US Institute of Peace, September 29, 2010, www.usip.org/publications/crowdsourcing-crisis-information-in-disaster -affected-haiti.

15. Steve Coast, interview with the author, November 1, 2009.

16. Ibid., January 25, 2012.

17. Roberto Rocha, "A Map-Making Democracy," *Montreal Gazette,* December 20, 2007.

18. Michael Cross, "OS Maps Finally Available to Not-for-Profit Organisations," *Guardian,* December 13, 2007.

19. Jonathan Brown, "No. 1 in the Charts Since 1747; Now Its Maps Are Available on the Web," *Independent,* April 2, 2010.

20. Carl Franzen, "OpenStreetMap Reaches 1 Million Users, Will Rival Google Maps in 2 Years," *Talking Points Memo,* January 12, 2013, http://idealab .talkingpointsmemo.com/2013/01/openstreetmap-reaches-1-million-users -will-rival-google-maps-in-2-years.php.

21. Rocha, "A Map-Making Democracy."

22. "MapQuest to Launch Open-Source Mapping in Europe," *Wireless News,* July 20, 2010.

23. Rob D. Young, "Google Maps API to Charge for High-Volume Usage," *Search Engine Watch,* November 2, 2011, http://searchenginewatch.com /article/2122151/Google-Maps-API-to-Charge-for-High-Volume-Usage.

24. Quentin Hardy, "Facing Fees, Some Sites Are Bypassing Google Maps," *New York Times,* March 20, 2012.

25. Google Geo Developers blog, June 22, 2012, http://googlegeodevelopers .blogspot.co.uk/2012/06/lower-pricing-and-simplified-limits.html.

26. Keynote speech by Ed Parsons, Google geospatial technologist, at Google PinPoint London conference, November 27, 2012, www.youtube. com/watch?v=ucYiMBfyNfo.

27. Dave Barth, "The Bright Side of Sitting in Traffic: Crowdsourcing Road Congestion Data," Google Official Blog, August 25, 2009, http://googleblog.blogspot.com/2009/08/bright-side-of-sitting-in-traffic.html.

28. Catherine Shu, "Nav App Waze Says 36M Users Shared 900M Reports, While 65K Users Made 500M Map Edits," *TechCrunch*, February 6, 2013, http://techcrunch.com/2013/02/06/nav-app-waze-says-36m-users-shared-900m-reports-while-65k-users-made-500m-map-edits/.

29. Jessica Guynn, "Google Acquisition Keeps Waze Out of Rivals' Hands," *Los Angeles Times*, June 12, 2013.

Chapter 9

1. John Perry Barlow, "A Declaration of the Independence of Cyberspace," February 8, 1996, https://projects.eff.org/~barlow/Declaration-Final.html.

2. Thomas Lowenthal, "IP Address Can Now Pin Down Your Location to Within a Half Mile," *Ars Technica*, April 22, 2011, http://arstechnica.com/tech-policy/2011/04/getting-warmer-an-ip-address-can-map-you-within-half-a-mile/.

3. Bobbie Johnson, "Money Can't Buy You Loyalty," *Guardian*, April 30, 2007.

4. Vindu Goel, "Why Google Pulls the Plug," *New York Times*, February 15, 2009.

5. Erick Schonfeld, "Dennis Crowley on the Origins of Foursquare," *TechCrunch*, March 2, 2011, http://techcrunch.com/2011/03/02/founder-stories-crowley-foursquare-origins/.

6. John Timpane, "You Can Be the Mayor—Anywhere; Foursquare Anoints Users Who've Arrived," *Philadelphia Inquirer*, July 17, 2011.

7. "On Foursquare, Cheating, and Claiming Mayorships from Your Couch," Foursquare Blog, April 7, 2010, http://blog.foursquare.com/2010/04/07/503822143/.

8. Matthew Flamm, "Foursquare Doesn't Quite Check Out," *Crain's New York Business*, January 21, 2013.

9. David Petersen, interview with the author, March 14, 2013.

10. Ken Yeung, "Life360, the Location App for Families, Nears 25M Users," *NextWeb*, December 18, 2012, http://thenextweb.com/insider/2012/12/18/life360-to-hit-25-million-users-adds-geo-fencing-feature/.

11. Alistair Goodman, interview with the author, March 13, 2013.

12. "Broadcom Introduces GNSS Location Chip with Geofence Capabilities," February 20, 2013, www.broadcom.com/press/release.php?id=s741713.

Chapter 10

1. Henri E. Cauvin, "Cash and Cocaine, but No Conviction," *Washington Post,* March 5, 2007.

2. Chaoming Song et al., "Limits of Predictability in Human Mobility," *Science,* February 19, 2010, 1018–1021.

3. Jeff Jonas, "Big Data, New Physics, and Geospatial Super-Food," presentation at GigaOM Structure Data conference, www.livestream.com /gigaombigdata/video?clipId=pla_f16e2bd9–24f7–4216-b52b-e5b3cf05 9a9b.

4. "Appthority App Reputation Report," February 2013, www.appthority .com/appreport.pdf.

5. Hiawatha Bray, "Smartphone Apps Track Users Even When Shut Down," *Boston Globe,* September 2, 2012.

6. "Acxiom Corp.: The Faceless Organization That Knows Everything About You," *Week,* June 20, 2012.

7. John Gilliom and Torin Monahan, *SuperVision: An Introduction to the Surveillance Society* (Chicago: University of Chicago Press, 2013), 20.

8. *US v. Jones,* 132 S. Ct. 945 (2012).

9. Brief of the United States, *US v. Jones,* 132 S. Ct. at 13.

10. US Code 18 USC § 2703.

11. Ann E. Marimow, "Judge Declares Mistrial in Latest Prosecution of Antoine Jones," *Washington Post,* March 5, 2013.

12. Jon Campbell, "LAPD Spy Device Taps Your Cell Phone," *LA Weekly,* September 13, 2012.

13. *United States v. Rigmaiden,* No. 08–814, 2012 WL 1038817.

14. Jennifer Valentino-DeVries, "Stingray Phone Tracker Fuels Constitutional Clash," *Wall Street Journal,* September 22, 2011; Kim Zetter, "Feds' Use of Fake Cell Tower: Did It Constitute a Search?," Wired.com, November 3, 2011, www.wired.com/threatlevel/2011/11/feds-fake-cell-phone-tower/.

15. Cynthia Lum and Linda M. Merola, "Emerging Surveillance Technologies: Privacy and the Case of License Plate Recognition (LPR) Technology," *Judicature* (November–December 2012): 119.

16. Jon Campbell, "License-Plate Recognition Has Its Eyes on You," *San Diego CityBeat,* February 20, 2013, www.sdcitybeat.com/sandiego/article -11511-license-plate-recognition-has-its-eyes-on-you.html.

17. "Report: RFID Market to Surpass $26B in 2022," *Security Sales and Integration,* July 19, 2012, www.securitysales.com/channel/vertical-markets /news/2012/07/19/report-rfid-market-to-surpass-26b-in-2022.aspx. See also

Joanne Perry, "RFID Developments," *Aircraft Technology* (December 2012–January 2013): 56–61.

18. Hal Abelson, Ken Ledeen, and Harry Lewis, *Blown to Bits: Your Life, Liberty, and Happiness After the Digital Explosion* (Upper Saddle River, NJ: Addison-Wesley Professional, 2008), 36–38.

19. Katherine Albrecht and Liz McIntyre, *Spychips* (Nashville: Thomas Nelson, 2005), 35.

20. Jerry Banks et al., *RFID Applied* (New York: John Wiley and Sons, 2007), 8–10.

21. Albrecht and McIntyre, *Spychips*, 177.

22. Torin Monahan and Jill A. Fisher, "Surveillance Impediments: Recognizing Obduracy with the Deployment of Hospital Information Systems," *Surveillance and Society* 9, nos. 1–2 (2001): 1–16, 2011.

23. Kayode Aladesuyi, interview with the author, April 15, 2013.

24. *Hernandez v. Northside Independent School District*, US District Court, Western District of Texas, SA-12-CA-1113-OG.

25. Will Oremus, "Texas School District Drops RFID Chips, Will Track Kids with Surveillance Cameras Instead," July 17, 2031, www.slate.com /blogs/future_tense/2013/07/17/texas_northside_school_district_drops_rfid _tracking_privacy_not_the_main.html.

26. "Consumers Against Supermarket Privacy Invasion and Numbering," *Position Paper on the Use of RFID in Schools*, www.spychips.com/school /RFIDSchoolPositionPaper.pdf.

27. Gilliom and Monahan, *SuperVision*, 78–79.

28. See Electronic Communications Privacy Act Amendments Act of 2013, S. 607, Online Communications and Geolocation Protection Act, H.R. 983; and Geolocation Privacy and Surveillance Act, H.R. 1312, S. 639.

29. Location Privacy Protection Act of 2012, S. 1223.

30. Andrew J. Blumberg and Peter Eckersley, "On Locational Privacy, and How to Avoid Losing It Forever," Electronic Frontier Foundation, August 2009, www.eff.org/wp/locational-privacy.

31. Andrew J. Blumberg, Lauren S. Keeler, and Abhi Shelat, "Automated Traffic Enforcement Which Respects Driver Privacy," *Proceedings of the 7th International IEEE Conference on Intelligent Transportation Systems* (September 2005): 650–655.

Index